PLAGUE OF CORRUPTION

PLAGUE OF CORRUPTION

RESTORING FAITH IN THE PROMISE OF SCIENCE

Dr. Judy Mikovits
& Kent Heckenlively, JD

FOREWORD BY ROBERT F. KENNEDY, JR.

Skyhorse Publishing

Skyhorse Publishing books may be purchased in bulk at special discounts for
sales promotion, corporate gifts, fund-raising, or educational purposes. Special
editions can also be created to specifications. For details, contact the Special Sales
Department, Skyhorse Publishing, 307 West 36th Street, 11th Floor, New York,
NY 10018 or info@skyhorsepublishing.com.

Skyhorse® and Skyhorse Publishing® are registered trademarks of Skyhorse
Publishing, Inc.®, a Delaware corporation.

Visit our website at www.skyhorsepublishing.com.

10 9 8 7 6 5 4

Library of Congress Cataloging-in-Publication Data is available on file.

Cover Design: Paul Qualcom

Print ISBN: 978-1-5107-5224-5
Ebook ISBN: 978-1-5107-5225-2

Printed in the United States of America

To the brave who always fight for the truth.

To go against conscience is neither right nor safe.
Therefore, I cannot and will not recant. Here I stand.
I can do no other. God help me, amen.

From the movie *Martin Luther* (1953)

Contents

Foreword by Robert F. Kennedy, Jr.   ix

Introduction by Dr. Judy Mikovits   xxi

1. A Scientist at Sea   1

2. A Rebel from the Start   19

3. The Dead Doctors—What Is Real?   31

4. The Fate of Those Who Fight the Darkness   49

5. Is the Government a Friend or Foe?   67

6. The Blood Working Group and the Cerus Boondoggle   77

7. VP62—The Clone Assassin   97

8. My Identity Stolen in Vaccine Court   111

9. What I Really Think about HIV and Ebola   141

10. My Coauthor Gets Banned from Australia   163

11. The Way Forward   179

12. One More Story I Should Probably Tell   203

Notes   211

Acknowledgments   219

Foreword

Moral Courage and Our Common Future

by Robert F. Kennedy, Jr.

"And yet, it moves!" Galileo whispered those defiant words in 1615 as he left the Roman Inquisition tribunal before which he repudiated his theory that the Earth—the immovable center of the Universe according to contemporary orthodoxy—revolves around the sun. Had he not recanted, his life would be forfeit. We like to think of Galileo's struggles as the quaint artifact of a dark, ignorant, and tyrannical era where individuals challenged government-anointed superstitions only at grave personal risk. Dr. Judy Mikovits' story shows that stubborn orthodoxies anointed by pharmaceutical companies and corrupt government regulators to protect power and profits remain a dominant force in science and politics.

By any standard, Dr. Judy Mikovits was among the most skilled scientists of her generation. She entered professional science from the University of Virginia with a BA degree in chemistry on June 10, 1980, as a protein chemist for the National Cancer Institute (NCI) working on a life-saving project to purify interferon. The quality of her work and her reliable flashes of genius soon propelled her to the apex of the male-dominated world of scientific research. At NCI, Mikovits began what would become a twenty-year collaboration with Dr. Frank Ruscetti, a pioneer in the field of human retrovirology. While heading up the lab of Robert Gallo in 1977, Ruscetti made scientific history by codiscovering with Bernie Poiesz the first human retrovirus, HTLV-1 (human T-cell leukemia virus). A retrovirus is a "stealth virus" that, like HIV, enters the host without alerting the immune systems.

It may then lie dormant for years without causing harm. Before killing a person, a retrovirus will usually destroy their immune system. As a result, many retroviruses cause cancer. With an escalating understanding of retrovirus behavior, the Ruscetti/ Mikovits collaboration and Mikovits's award-winning PhD thesis from George Washington University in 1991 changed the paradigm of HIV-AIDS treatment, turning the disease from a death sentence into a manageable condition.

From the outset, the most daunting obstacle to Mikovits' career advancement was her scientific integrity. She always placed it ahead of personal ambition. Judy Mikovits never meant to wade into a public health brawl. She never considered herself a renegade or revolutionary. Judy's relatives mainly worked in government or law enforcement. They believed in the bedrock American principles of hard work, respect for authority, and, above all, telling the truth. That backdrop made it impossible for her to abandon her high natal standards of honesty and integrity even when they became a hindrance.

After leaving NIH, she worked a stint for Upjohn—leading a project to prove the safety of the company's blockbuster Bovine Growth Hormone. When Mikovits discovered the company's formula could cause precancerous changes in human cell cultures, she refused direct orders from her boss to hide her discoveries. Mikovits' revelation suggested that the ubiquitous presence of the hormone in milk could lead to breast cancer in women who drank it. Her refusal to back down precipitated her departure from Upjohn and her return to NIH and graduate school. Judy's war on BGH eventually led to Upjohn abandoning the product.

In 2009, now in academia, Mikovits and Ruscetti, who was still at NCI, led a team that discovered a strong association between a previously unknown retrovirus and myalgic encephalomyelitis, commonly known as chronic fatigue syndrome (ME/CFS). Predictably, the retrovirus was also linked to certain blood cancers. Collaborators had named it Xenotropic Murine Leukemia Related Virus (XMRV), when they first detected it in DNA sequences in prostate cancer a few years earlier.

The medical community had dealt with Chronic Fatigue Syndrome, which strikes mostly women, in bad faith since its appearance in the mid-1980s. The medical establishment derided ME/CFS as "yuppie flu" and attributed it to the inherent psychological fragility of career women pursuing professions in high-pressure corporate ecosystems. Mikovits found evidence for the retrovirus in approximately 67 percent of women afflicted with ME/CFS, and in a little less than 4 percent of the healthy population.

On October 8, 2009, Mikovits and Ruscetti published their explosive findings in the journal *Science*, describing the first-ever isolation of the recently discovered retrovirus XMRV, and its association to ME/CFS. Her revelation about ME/CFS immediately triggered angry reactions from jealous cancer power centers, stubbornly resistant to science that attributed cancer and neuroimmune diseases to viruses.

The blowback grew even grimmer when Mikovits' subsequent research suggested that the new retrovirus, originally found in mice, had somehow jumped into humans via contaminated vaccines.

Even more troubling to the medical establishment, Dr. Mikovits' research revealed that many of the female patients afflicted with XMRV had children with autism. Suspecting XMRV might be passed from mother to child, as with HIV, Mikovits tested seventeen of the children. Fourteen showed evidence of the virus. Those findings dovetailed with parental reports of autistic regression following vaccination. Subsequent studies linked XMRV to epidemics in leukemia, prostate cancer, autoimmune disease, and the explosion of Alzheimer's disease.

Worse yet, research also found widespread XMRV contamination in the blood supply and blood products. Based on her research and the findings of others, it seemed that anywhere from 3 to 8 percent of the population now carry the virus—XMRV has become part of human ecology, passed from mother to child in vitro or through breast milk. Mikovits' data suggest that more than ten million Americans are harboring this virus like a ticking time bomb—a potential threat far greater than the HIV-AIDS epidemic.

In January of 2011, HIV-AIDS expert Ben Berkhout published these explosive revelations in the journal *Frontiers in Microbiology*. He included Mikovits' evidence that mouse tissue used in vaccine production was the likely vector for human contamination. Unbeknownst to Judy, her co-author on this book, Kent Heckenlively, had already independently discovered published medical research showing that the first recorded outbreak of ME/CFS was among 198 doctors and nurses at the Los Angeles County Hospital in 1934–1935, following their injection with an experimental polio vaccine grown in mouse brain tissue.

Mikovits' evidence threatened financial catastrophe for the world's pharmaceutical companies because of their negligent use of animal cell cultures to produce vaccines and other pharmaceutical products. Her findings put at risk billions of dollars of revenues from an entire branch of medicine called "biologics," which depends on animal tissue and products.

Pharmaceutical companies and their captive regulators unleashed a furious broadside against Mikovits and Ruscetti, besieging them from every stronghold.

The journal *Science* feverishly pressed Mikovits to retract her October 2009 article. In September of 2011, the Whittemore Peterson Institute at the University of Nevada, Reno, fired Judy from her faculty job. Judy and her family noticed menacing-looking men following her in pickup trucks and other incidents indicating she was under surveillance. In one incident, burley thugs surrounded her home and forced her to flee in a boat. After she escaped, they barged into her home, claiming to work for the government. In November, Ventura Police arrested Judy without a warrant and held her in jail for five days without bail. The police searched her house from top to bottom, strewing her papers everywhere. That same day, cops raided the home of her friend, Lilly, and forced her to sit in a chair for several hours while they ransacked the building. NIH officials told Nevada police that Dr. Mikovits had illegally taken her research notebooks from their lab. This was a fabricated charge. As the principal investigator on two government grants, it was Dr. Mikovits' obligation to retain all of her research papers . . . Furthermore, Judy had left all of the notebooks in her university office on September 29. That same day, someone illegally burglarized Judy's office, removed her notebooks, and then somehow planted them in a closet of her home, apparently to incriminate her. Weeks later, as Judy languished in a cell, her husband, David, found the journals neatly packed in a linen beach bag in an obscure closet in her Southern California home. David frantically took them to the jail after midnight and then handed them over to Ventura Police.

While she was in jail, Judy's former boss told her husband and Dr. Ruscetti that if she just signed an apology admitting her paper was wrong, the police would release her from confinement and she could salvage her science career. Judy refused. No prosecutor has ever filed charges against her, but the pharmaceutical cartel and its captive scientific journals launched a campaign of vilification against her. Less than two years earlier, the journal *Science* had celebrated her. Now, the same journal published her mug shot and retracted her paper.

Judy lost federal grants for which she was the principal investigator. She has gone bankrupt trying to find work and restore her good name. The scientific journals, admittedly all now controlled by Big Pharma, have refused to publish her papers. The NIH medical libraries have locked her out. Despite spending hundreds of thousands of dollars in legal fees, she has not been able to get her day in court. The US Attorney in Nevada has kept the case "under seal" for years. Fraudulent acts of public health officials at the highest levels of Health and Human Services (HHS) have effectively rendered her unemployable.

The persecution of scientists and doctors who dare to challenge contemporary orthodoxies did not take a rest after Galileo: it has always been, and remains today, an occupational hazard. Henrik Ibsen's 1882 play *An Enemy of the People* is a parable for the pitfall of scientific integrity. Ibsen tells the story of a doctor in southern Norway who discovers that his town's popular and lucrative public baths were actually sickening the visitors who flocked to them for rejuvenation. Discharges from local tanneries had infected the spas with lethal bacteria. When the doctor goes public with the information, local merchants, joined by government officials, their allies in the "liberal-minded independent press," and other financially interested parties move to muzzle him. The medical establishment pulls his medical license, the townsfolk vilify and brand him "an enemy of the people."

Ibsen's fictional doctor experienced what social scientists call the "Semmelweis reflex." This term describes the knee-jerk revulsion with which the press, the medical and scientific community, and allied financial interests greet new scientific evidence that contradicts an established scientific paradigm. The reflex can be particularly fierce in cases where new scientific information suggests that established medical practices are actually harming public health.

The real-life plight of Ignaz Semmelweis, a Hungarian physician, inspired the term and Ibsen's play. In 1847, Dr. Semmelweis was an assistant professor at Vienna's General Hospital maternity clinic, where around 10 percent of women died from puerperal "birth bed" fever. Based on his pet theory that cleanliness could mitigate transmission of disease-causing "particles," Semmelweis introduced the practice of mandatory hand washing for interns between performing autopsies and delivering babies. The rate of fatal puerperal fever immediately dropped to around 1 percent. Semmelweis published these findings.

Rather than building a statue to Semmelweis, the medical community, unwilling to admit culpability in the injury of so many patients, expelled the doctor from the medical profession. His former colleagues tricked Dr. Semmelweis into visiting a mental institution in 1865, then committed him against his will. Semmelweis died mysteriously two weeks later. A decade afterward, Louis Pasteur's germ theory and Joseph Lister's work on hospital sanitation vindicated Semmelweis's ideas.

Modern analogs abound. Herbert Needleman of the University of Pittsburgh endured the Semmelweis reflex when he revealed the brain-killing toxicity of lead in the 1980s. Needleman published a groundbreaking study in 1979 in the *New England Journal of Medicine* showing that children with high levels of lead in their teeth scored significantly lower than their peers on intelligence tests, on auditory and speech processing, and on attention measurements. Beginning in the early 1980s, the lead and oil industries (leaded gasoline was a lucrative petroleum product) mobilized public relations firms and scientific and medical consultants to lambast Needleman's research and his credibility. Industry pressured the Environmental Protection Agency, the Office of Scientific Integrity at the National Institutes of Health, and the University of Pittsburgh to launch investigations against Needleman. Ultimately the federal government and the University vindicated Needleman. But the impact of the industry's scathing assault ruined Needleman's academic career and stagnated the field of lead research. The episode offered an enduring demonstration of industry power to disrupt the lives of researchers who dare to question their products' safety.

Rachel Carson ran the same gauntlet in the early 1960s when she exposed the dangers of Monsanto's DDT pesticide, which the medical community then promoted as prophylactic against body lice and malaria. Government officials and medical professionals led by the American Medical Association joined Monsanto and other chemical manufacturers, attacking Carson viciously. Trade journals and the popular media disparaged her as a "hysterical woman." Industry talking points derided Carson as a "spinster," the contemporary euphemism for lesbian, and for being unscientific. Vicious criticisms of her book appeared in editorial pages in *Time, Life, Newsweek,* the *Saturday Evening Post, US News and World Report,* and even *Sports Illustrated.* I am immensely proud that my uncle, President John F. Kennedy, played a critical role in vindicating Carson. In 1962, he defied his own USDA, a captive agency in league with Monsanto, and appointed a panel of independent scientists who validated every material assertion in Carson's book *Silent Spring.*

The experience of British physician and epidemiologist Alice Stewart offers a near-perfect analogy to the Medical cartel's lynching of Judy Mikovits. In the 1940s, Stewart was one of the rare women in her profession and the youngest fellow ever elected at the time to the Royal College of Physicians. She began investigating the high occurrences of childhood cancers in well-to-do families, a puzzling phenomenon given that disease often correlated with poverty, and seldom with affluence. Stewart published a paper in *The Lancet* in 1956 offering strong evidence that the common practice of giving X-rays to pregnant women was the culprit in carcinomas that would later afflict their children. According to Margaret Heffernan, author of *Willful Blindness*, Stewart's finding "flew in the face of conventional wisdom"—the medical profession's enthusiasm for the new technology of X-rays—as well as "doctors' idea of themselves, which was as people who helped patients." A coalition of government regulators, nuclear promoters, and the nuclear industry joined the US and British medical establishments in launching a brutal attack on Stewart. Stewart, who died in 2002 at the age of ninety-five, never again received another major research grant in England. It took twenty-five years after the publication of Stewart's paper for the medical establishment to finally acknowledge her findings and abandon the practice of X-raying expectant mothers.

Judy Mikovits is heir to these martyrs and, more directly, to a long line of scientists, whom public health officials have punished, exiled, and ruined specifically for committing heresy against reigning vaccine orthodoxies.

Dr. Bernice Eddy was an award-winning virologist, and one of the highest ranking female scientists in NIH history. She and her research partner Elizabeth Stewart were the first researchers to isolate the Polyomavirus—the first virus proven to cause cancer. In 1954, NIH asked Eddy to direct testing of the Salk polio vaccine. She discovered, while testing eighteen macaques, that Salk's vaccine contained residual live polio virus that was paralyzing the monkeys. Dr. Eddy warned her NIH bosses that the vaccine was virulent, but they dismissed her concerns. The distribution of that vaccine by Cutter Labs in California caused the worst polio outbreak in history. Health officials infected 200,000 people with live polio; 70,000 became sick, leaving 200 children paralyzed and ten dead.

In 1961, Eddy discovered that a cancer-causing monkey virus, SV40, had contaminated ninety-eight million Salk polio vaccines. When she injected the SV40 virus into newborn hamsters, the rodents sprouted tumors. Eddy's discovery proved an embarrassment to many scientists working on the vaccine. Instead of rewarding her for her visionary work, NIH officials banned her from polio research and assigned her to other duties. The NIH buried the alarming information and continued using the vaccines.

In the autumn of 1960, the New York Cancer Society invited Eddy to address its annual conference. Eddy chose the subject of tumors induced by the polyoma virus. However, she also described tumors induced by the SV40 viral agent in monkey kidney cells. Her NIH supervisor angrily reprimanded Eddy for mentioning the discovery publicly and banned her from public health crisis statements. Eddy argued for publication of her work on the virus, casting the contaminated vaccine supply on an urgent public health crisis. Agency bigwigs stonewalled publication, allowing Merck and Parke-Davis to continue marketing the oncogenic vaccine to millions of American adults and children.

On July 26, 1961, the *New York Times* reported that Merck and Parke-Davis were withdrawing their Salk vaccines. The article said nothing about cancer. The *Times* ran the story next to an account about overdue library fines on page 33.

Even though Merck and Parke-Davis recalled their polio vaccines in 1961, NIH officials refused to pursue a total recall of the rest of the supply, fearing reputational injury to the vaccine program if Americans learned that PHS had infected them with a cancer-producing virus. As a result, millions of unsuspecting Americans received carcinogenic vaccines between 1961 and 1963. The Public Health Service then concealed that "secret" for forty years.

In total, ninety-eight million Americans received shots potentially containing the cancer-producing virus, which is now part of the human genome. In 1996, government researchers identified SV-40 in 23 percent of the blood specimens and 45 percent of the sperm specimens collected from healthy adults. Six percent of the children born between 1980 and 1995 are infected. Public health officials gave millions of people the vaccine for years after they knew it was infected. They contaminated humanity with a monkey virus and refused to admit what they'd done.

Today, SV-40 is used in research laboratories throughout the world because it is so reliably carcinogenic. Researchers use it to produce a wide variety of bone and soft-tissue cancers including mesothelioma and brain tumors in animals. These cancers have exploded in the baby boom generation, which received the Salk and Sabin polio vaccines between 1955 and 1963. Skin cancers are up by 70 percent, lymphoma and prostate by 66 percent, and brain cancer by 34 percent. Prior to 1950, mesothelioma was rare in humans. Today, doctors diagnose nearly 3,000 Americans with mesotheliomas every year; 60 percent of the tumors that were tested contained SV-40. Today, scientists find SV-40 in a wide range of deadly tumors, including between 33 percent and 90 percent of brain tumors, eight of eight ependymomas, and nearly half of the bone tumors tested.

In successive measures, NIH forbade Bernice Eddy from speaking publicly or attending scholarly conferences, held up her papers, removed her from vaccine research altogether, and eventually destroyed her animals and took away access to her labs. Her treatment continues to mark an enduring scandal with the scientific community, yet NIH's Bernice Eddy playbook has become a standardized template for Federal vaccine regulators in their treatment of dissident vaccine scientists who seek to tell the truth about vaccines.

Dr. John Anthony Morris was a bacteriologist and virologist who worked for thirty-six years at NIH and the Food and Drug Administration (FDA), beginning in 1940. Morris served as the chief vaccine officer for the Bureau of Biological Standards (BBS) at the National Institute of Health and later with the FDA when the BBS transferred to that agency in the 1970s. Dr. Morris irked his superiors by arguing that the research carried out by his unit demonstrated there was no reliable proof that flu vaccines were effective in preventing influenza; in particular, he accused his supervisor of basing HHS's mass vaccination program for the swine flu primarily on a scientifically baseless fear campaign and on false claims made by pharmaceutical manufacturers. He warned that the vaccine was dangerous and could induce neurological injuries. His CDC superior warned Dr. Morris, "I would advise you not to talk about this."

When vaccine recipients began reporting adverse reactions, including Guillain-Barré, Dr. Morris disobeyed that order and went public. He declared that the flu vaccine was ineffective and potentially dangerous and said that he could find no evidence that this swine flu was dangerous or that it would spread from human to human.

In retaliation, FDA officials confiscated his research materials, changed the locks on his laboratory, reassigned his laboratory staff, and blocked his efforts to publish his findings. The FDA assigned Dr. Morris to a small room with no telephone. Anyone who wished to see him had to secure permission from the chief of the lab. In 1976, HHS fired Dr. Morris on the pretext that he failed to return library books on time.

Subsequent events supported Dr. Morris's skepticism about the swine flu shot. The 1976 swine flu vaccination program was so fraught with problems that the government discontinued inoculations after forty-nine million people had received the vaccine. Among the vaccine's victims were 500 cases of Guillain-Barré, including 200 people paralyzed and thirty-three dead. Furthermore, the incidence of swine flu among vaccinated was seven times greater than among those who were unvaccinated, according to news reports.

According to his *New York Times* obituary, Dr. Morris said, "The producers of these (influenza) vaccines know they are worthless, but they go on selling them anyway." He told the *Washington Post* in 1979, "It's a medical ripoff. . . . I believe the public should have truthful information on the basis of which they can determine whether or not to take the vaccine," adding, "I believe that given full information, they won't take the vaccine."

FDA used the same playbook in 2002 to isolate, silence, and drive from government service its star epidemiologist, Dr. Bart Classen, when his massive epidemiologic studies, the largest ever performed, linked Hib vaccines to the juvenile diabetes epidemic. FDA ordered Dr. Classen to refrain from publishing the government-funded studies, forbade him from talking publicly about the alarming outbreak, and eventually forced him out of government service.

In 1995, the CDC hired a PhD computer analytics expert, Dr. Gary Goldman, to perform the largest-ever CDC-funded study of the chickenpox vaccine. Goldman's results on an isolated population of 300,000 residents of Antelope Valley, California, showed that the vaccine waned, leading to dangerous outbreaks of chickenpox in adults and that ten-year-old children who received the vaccine were getting shingles at over three times the rate of unvaccinated children. Shingles has twenty times the death rate of chickenpox and causes blindness. CDC ordered Goldman to hide his findings and forbade him from publishing his data. In 2002, Goldman resigned in protest. He sent a letter to his bosses saying that he was resigning because "I refuse to participate in research fraud."

Recent medical history overflows with other examples of the brutal suppression of any science that exposes vaccines' risks; its casualties include brilliant and compassionate doctors and scientists like Dr. Waney Squier, the railroaded British gastroenterologist Andy Wakefield, the steadfast father/son research team David and Dr. Mark Geier, and Italian biochemist Antionetta Gatti. Any just society would have built statues to these visionaries and honored them with laurels and leadership. Our corrupt medical officials have systematically disgraced and silenced them.

In England a neuropathologist, Dr. Waney Squier of the Radcliffe Hospital in Oxford, testified in a series of cases on behalf of defendants accused of inflicting shaken baby syndrome. Squier believed that, in these cases, vaccines and not physical trauma had caused the infants' brain injuries. In March 2016, the Medical Practitioner's Tribunal Service (MPTS) charged her with falsifying evidence and lying and struck her from the medical register. Squier appealed the tribunal's decision in November 2016. The High Court of England reversed the MPTS's decision, concluding, "The determination of the MPTS is in many significant ways flawed."

Professor Peter Gøtzsche cofounded the Cochrane Collaboration in 1993 to remedy the overwhelming corruption of published science and scientists by pharmaceutical companies. Over 30,000 of the world's leading scientists joined Cochrane as volunteer reviewers hoping to restore independence and integrity to published science. Gøtzsche was responsible for making Cochrane the world's leading independent research institute. He also founded the Nordic Cochrane Center in 2003. On October 29, 2018, pharmaceutical interests, led by Bill Gates, finally succeeded in ousting Professor Gøtzsche. A stacked board controlled by Gates fired Gøtzsche from the Cochrane Collaboration after he published a well-founded criticism of the HPV vaccine. In 2018, the Danish government, under pressure from pharma, fired Peter Gøtzsche from Rigshospitalet in Copenhagen. His findings about the HPV vaccine threatened the pharmaceutical industry's earnings.

Science, at its best, is a search for existential truth. Sometimes, however, those truths threaten powerful economic paradigms. Both science and democracy rely on the free flow of accurate information. Greedy corporations and captive government regulators have consistently shown themselves willing to twist, distort, falsify, and corrupt science, hide information, and censor open debate to protect personal power and corporate profits. Censorship is the fatal enemy of both democracy and public health. Dr. Frank Ruscetti often quotes Valery Legasov, the courageous Russian physicist who braved censor, torture, and threats on his life by the KGB to reveal to the world the true cause of the Chernobyl disaster. "To be a scientist is to be naïve. We are so focused on our search for the truth, we fail to consider how few actually want us to find it. But it is always there, whether we can see it or not, whether we choose to or not. The truth doesn't care about our needs or our wants. It doesn't care about our governments, our ideologies, our religions. It will lie in wait for all time."

This account by Judy Mikovits and Kent Heckenlively is vitally important both to the health of our children and the vitality of our democracy. My father believed moral courage to be the rarest species of bravery. Rarer even than the physical courage of soldiers in battle or great intelligence. He thought it the one vital quality required to salvage the world.

If we are to continue to enjoy democracy and protect our children from the forces that seek to commoditize humanity, then we need courageous scientists like Judy Mikovits who are willing to speak truth to power, even at terrible personal cost.

Introduction

by Dr. Judy Mikovits

I never imagined I'd become one of the most controversial figures in twenty-first century science.

For me it was always about following the data and listening to the patients. As scientists we're supposed to solve problems and help humanity. That's our mission, the very purpose to which we've dedicated our lives.

How is it that conditions such as chronic fatigue syndrome (CFS), autism, neurological diseases, and even cancer have become so controversial that many in medicine turn away from looking at the possible root causes of these diseases?

I don't know why we can't all just put our heads together and figure this out. Maybe some of my ideas are right and maybe some are wrong. Let's put all the ideas under the microscope and see what happens.

When I was in a lab, making breakthrough discoveries, such as how to develop AIDS drugs to solve our world's greatest modern plague, the HIV-AIDS epidemic, it was never about glory or reputation.

It was about changing people's lives.

That's where I derive my greatest satisfaction.

I don't consider myself one who seeks the limelight. When I worked in a lab, I was usually in by five in the morning, often not leaving until six that night (unless there was a baseball game at Camden Yards). I read a lot of scientific articles, trying to understand what the best minds in the field are finding. When I relax, I like to watch baseball, which explains why I often wear a baseball cap. It's just more comfortable to me. I like other sports as well, basketball and football, and for many years I've belonged to what we jokingly call the "Poor Boys Yacht Club" (actually, it's the Pierpont Bay

Yacht Club, but we prefer PBYC) and enjoyed sailing on the Pacific Ocean with my husband and our friends. I participate in cancer support groups, using my knowledge as a researcher to help people going through therapy by explaining options their doctors suggest.

None of that explains why the police raided my house on a mid-November morning in 2011 and held me in jail without bail for five days. I didn't murder anybody. I'm not the agent of a foreign power. In fact, I was never even tried for a crime, the allegation vanishing like an early morning California fog.

What do I think was my real crime?

Following the data and listening to the patients.

We think the key to solving these questions really starts in 1934 in sunny Los Angeles, the City of Angels.

* * *

Most experts agree that the first appearance of CFS in the United States occurred in Los Angeles between 1934 and 1935. The outbreak affected 198 doctors, nurses, medical technicians, and other workers at the Los Angeles County Hospital during a polio outbreak.

Oddly enough, only hospital staff contracted CFS, and the patients managed to avoid it.

Doesn't that sound like a clue to you?

The signs and symptoms were puzzling. The patients were easily fatigued upon the slightest exertion, had nausea, light sensitivity, loss of balance, episodes of paralysis followed by difficulty in lifting their limbs, breathing problems, headaches, shooting pains, and insomnia, in addition to difficulties with concentration and memory. The victims suffered from crushing depression followed by euphoria, spells of weeping for no apparent reason, and often showed violent manifestations of dislike for people or things they had previously liked.[1] It was as if their bodies and brains had betrayed them.

I entered the picture in May of 2006, when I heard a lecture from a researcher who noted that long-time sufferers of CFS had elevated rates of very rare types of cancer. It smelled like a virus to me, the same way HIV (human immunodeficiency virus) often kills its victims through the accompanying AIDS (acquired immunodeficiency syndrome) and the various cancers and other problems that their immune system could no longer handle.

The HIV-AIDS epidemic, which ravaged the planet in the 1980s and 1990s, killing more than thirty-five million people by the latest count, serves as an important counterpoint to the CFS epidemic. CFS seemed to explode in the 1980s, starting with an outbreak in Lake Tahoe on the California/Nevada border from 1984 to 1985, striking first the teachers at a local high school, then moving into the more urban areas of San Francisco, Los Angeles, and New York. From these locations it spread across the country. At that time, the dogma was HIV-AIDS only affected men, so CFS was often called non-HIV-AIDS.

HIV-AIDS killed its patients over the course of several years. CFS kept its victims alive, but in a state akin to hibernation. Friends and family members would often tell the patients they "looked great" and maybe just "needed to get out more" and try to "reduce their stress." Many of the same immune markers were abnormal in both groups, but the outcomes were very different.

Those with CFS remained alive but often longed for the release of death.

* * *

In science the first outbreak of a disease is usually closely examined for possible clues.

The researcher asks, what factors do those with the disease share? I'm sure you've seen that same protocol followed in countless books, movies, and television shows.

The same thing happened, at least initially, with that first outbreak of CFS among the hospital staff in Los Angeles in 1934–1935. The investigators who happened to be on hand when this new disease first appeared were Dr. John R. Paul, a Yale Medical School professor, and Dr. Leslie Webster, a physician with the Rockefeller Institute of New York.

In a book published in 1971 about the history of polio, Dr. Paul devoted an entire chapter to this new disease, which came to be known as CFS but earlier was called myalgic encephalomyelitis (ME). ME literally means inflammation of the brain and spinal cord. Many people refer to this disease as chronic fatigue syndrome/myalgic encephalomyelitis or CFS/ME. Others think that it should be more accurately referred to as ME/CFS. Fatigue is a symptom of many diseases, and I think the use of this term has hindered the science and marginalized the patients for four decades. The community of patients prefers the term ME/CFS, and I will use it in this book.

Even decades later, Dr. Paul seemed to be haunted by what he had seen during this first outbreak of ME/CFS and its possible origin. Had they missed something of critical importance? Paul wrote:

Nonetheless the Los Angeles episode is a reminder that even those who believe themselves to be experts occasionally ride for a fall, although they may be extremely loathe to admit it, especially to their patients. It is sometimes the bitterest pill they have to swallow. The members of our team of investigators had somehow failed to recognize the treachery of the situation and had not emphasized sufficiently the possibility of a hysterical element or the intrusion of a polio-like illness on the scene. As a weak excuse, it may be said that we had our hearts so set on isolating poliovirus that we could think of little else.[2]

I often find myself longing for the honesty and self-reflection of researchers like Dr. Paul, although I suspect he was also keeping a few secrets of his own. As I reflect on the controversy our work has generated, I wonder what Paul meant by "the treachery of the situation." How could the search for the truth about a disease involve treachery?

Surely a virus does not care what scientists discover about it.

The question remains, what could science possibly be hiding?

* * *

Kent Heckenlively, the coauthor of my previous book, *PLAGUE*, and with whom I worked again on this book, found a possible answer.

Kent discovered published medical research indicating that the entire medical staff at the Los Angeles County General Hospital had received an early polio vaccine developed by Dr. Maurice Brodie, in collaboration with the Rockefeller Institute of New York.[3] The polio virus was passaged multiple times through mouse brain tissue. The use of mouse tissue to grow viruses was new in the 1930s, only previously having been used in the development of a Yellow Fever vaccine.[4] In addition, the medical staff was given an accompanying immune system booster, preserved with thimerosal, a mercury derivative.[5]

Kent also uncovered the transcript of a lecture given by Dr. G. Stuart about problems with the Yellow Fever vaccine and its mouse components to a gathering of the World Health Organization in 1953 in Uganda:

[T]wo main objections to this vaccine have been voiced, because of the possibility that: (i) the mouse brains employed in its preparations may be contaminated with a virus pathogenic for man although latent in mice . . . or may be the cause of demyelinating encephalomyelitis; (ii) the use, as antigen,

of a virus with enhanced neurotropic properties may be followed by serious
reactions involving the central nervous system.[6]

I remind you that the scientific name for CFS is myalgic encephalomyelitis.
A "virus with enhanced neurotropic properties" that could cause "serious
reactions involving the central nervous system"? Sounds like the plot of a
terrifying science-fiction movie.

But this was a presentation made to the World Health Organization!

"Is this how it could have happened?" Kent asked. "By using animal
tissue to grow viruses they were picking up other viruses from those animals
and injecting them into human beings as passengers in vaccines?"

I could only reply that it was a good question. I consider it acknowledged science that the passage of human viruses through different animal
species, such as mice, rabbits, dogs, or monkeys, often results in a less pathogenic virus, which might be used in a vaccine. However, the question of
whether other animal viruses were hitching a ride in that biological material
was less clear.

I asked my long-time collaborator and mentor, Dr. Frank Ruscetti, what
he thought of this question. He replied he'd asked the same question as a
young researcher and been told that the human immune system was superior to any animal viruses that might hitch a ride along with the vaccine.
He was also told by John Coffin, a few years older than Frank, in the tone
of a know-it-all older brother, "Don't bother to look for human retroviruses.
They don't exist."

Frank said his first reaction was "that's preposterous." Even from the
start, John Coffin was a fountain of bad ideas and misguided advice. I
would have my own fights with Coffin and share Frank's opinion.

Frank went on to discover the first-known disease-causing human retrovirus, HTLV-1, (human T-cell leukemia virus), along with Robert Gallo
and Bernie Poiesz, and the field of human retrovirology was born. Like most
engineers of disaster throughout history, Coffin was "often wrong, but never
in doubt."

* * *

Did researchers in the 1930s understand what they might have done?

In her book *Osler's Web*, detailing the course of the ME/CFS outbreak
starting in the mid-1980s, journalist Hillary Johnson recounted how she
had been told by a Canadian researcher that the 198 victims of the initial

outbreak in Los Angeles of 1934–1935 had received a settlement of approximately six million dollars. This settlement was supposedly made somewhere around 1939.[7]

Kent did some investigation as to who might have made such a payment in 1939, which in today's dollars would amount to more than a hundred million dollars. Who had that kind of money during the Great Depression? Kent suspected the Rockefeller Institute, since they had partially funded the first use of mouse tissue for vaccines. He also found a curious pattern from their public financial reports. In 1935, the Rockefeller Foundation reported assets of over a hundred and fifty-three million dollars. But after 1939 it had shrunk to just over a hundred and forty-six million dollars, a loss of more than seven million dollars.[8] Variations in the years before and after this period tended to be less than fifty thousand dollars.

This is circumstantial evidence, but it could potentially explain why there was so little scientific follow-up in the medical literature for the initial victims of the Los Angeles outbreak.

* * *

We knew none of this when, on October 8, 2009, we published explosive findings in the journal *Science*, describing the first-ever isolation of a recently discovered retrovirus, XMRV (Xenotropic Murine Leukemia Virus-Related Virus) and its association to ME/CFS.[9] We found evidence for the retrovirus in approximately 67 percent of those afflicted with ME/CFS, and in a little less than 4 percent of the healthy population.

While this was welcome news to those suffering from ME/CFS, it also meant that more than ten million Americans were harboring this virus like a ticking time bomb. What might awaken this virus in human beings and cause disease?

We suspected immune activation, since retroviruses like to hide out in the monocytes, B and T cells of the immune system. Because of our previous research in HIV-AIDS, we knew that the standard of care for children born to HIV-infected mothers was that their babies be immediately put on antiretroviral drugs prior to any vaccination. The very act of immune stimulation by a vaccine was likely to cause the HIV virus to replicate out of the control of the immune system and cause the fatal consequence of AIDS.

There's one important point I need to make about HIV among retroviruses. For those who were diagnosed with HIV in the 1980s, it meant a death sentence. Those who survived had a unique genetic profile, which we came to call "elite controllers." Most retroviruses do not kill in the rampant fashion of HIV. They cause immune disruption and lead to a vast array of diseases, including cancer. That is the challenge of this struggle. We need to stop the viruses that are disabling our population, robbing people of the quality of their lives and, only after years of torture, mercifully ending their existence.

We were interested in the pattern of disease among the families of those with ME/CFS. If XMRV was a virus that behaved similarly to HIV, it would tend to get passed down from mother to child. Early in our research we'd noted several children with autism born to the affected mothers and tested seventeen of these children for XMRV. Was autism nothing more than ME/CFS in the young, when their development requires massive amounts of energy to develop the neurological connections necessary for speech, social interaction, and organized thinking?

Fourteen of the seventeen children with autism tested positive for evidence of XMRV.

The findings dovetailed with parental reports of autistic regression after a vaccination, and we felt it should be publicly discussed, especially with the lessons learned from Ryan White, a child who contracted HIV from a blood transfusion. At the time, we had not considered the possibility that XMRV might have originally come from animal tissue used in the vaccines.

But the simple act of providing support to the autism community and their concerns about vaccinations was akin to treason for many of my scientific colleagues. Honestly, we tried to downplay that implication of our research.

But we weren't going to hide from it. Frank and I had seen firsthand the carnage that resulted from the dogma surrounding HIV-AIDS. We were not going to let the band play merrily along again while millions suffered and died.

That's never been our style, and it's decidedly not good science.

* * *

An article in the journal *Frontiers in Microbiology* published in January of 2011 placed the issue in stark terms:

> One of the most widely distributed biological products that frequently involved mice or mouse tissue, at least up to recent years, are vaccines, especially vaccines against viruses . . . It is possible that XMRV particles were present in virus stocks cultured in mice or mouse cells for vaccine production, and that the virus was transferred to the human population by vaccination.[10]

Does it make sense now why we weren't the most popular folks in the scientific community? Was it possible that scientists in the lab had made terrible mistakes decades earlier, putting the health of humanity at risk? Our research seemed to suggest that possibility.

When it became clear that our research was opening old wounds and the asking of inconvenient questions, a campaign of unprecedented ferocity was launched against us. Most of that story is covered in our previous book, *PLAGUE*. By the end of 2012, our work had been thoroughly discredited in the scientific community. I'd been arrested and jailed for five days and rendered unemployable by the fraudulent acts of those at the highest levels of Health and Human Services (HHS).

If you read the *Wikipedia* version of my life, you will find that our work has been discredited, that what we believed to be an infection was simply lab contamination, and for good measure you might also find the mug shot of me that was published in the journal *Science* when I was arrested, but curiously, not charged for supposedly "stealing" my own research journals, a requirement as the principal investigator on two large government grants and required by federal law. The principal investigator on a government grant is responsible for the security of all materials on the project. As of this date, more than seven years later, I have not been provided copies of a single page of my notebooks or those of my research team.

If I am a criminal, why were no charges ever brought against me? My record is clean. And in the years since my false arrest and imprisonment, why have I been unable to have a single day in court for a judge and jury to hear my claims, even though I have never given up the effort to receive process?

* * *

In September of 2013, Dr. Ian Lipkin, the man who the previous year had supposedly debunked our findings of a retroviral connection to ME/CFS, held an unusual public conference call. He had done further research with Dr. Jose Montoya of Stanford University. Using a cohort of patients very similar to the one we used for the *Science* paper (and the very cohort excluded from the 2012 study by Tony Fauci, head of the National Institute for Allergy and Infectious Diseases), Lipkin said:

> We found retroviruses in 85 percent of the sample pools. Again, it is very difficult at this point to know whether or not this is clinically significant. And given the previous experience with retroviruses in chronic fatigue, I am

going to be very clear in telling you, although I am reporting this as present in Professor Montoya's samples, neither he nor we have concluded that there is a relationship to disease.[11]

Got it? Even though he found retroviruses in 85 percent of the samples of the sick patients, and in *only* 6 percent of the controls, he can't figure out if it means anything. And more shockingly, they're not going to do any further investigation.

This is the nightmare of censored and "dangerous" science today. There are no data because the appropriate studies are forbidden and censored.

* * *

Before I get too far into this story, I need to spend a little time talking about my longtime colleague, Frank Ruscetti. Everything I am as a scientist I owe to him.

I often compare our two personalities to Thomas Jefferson and Alexander Hamilton. These two men have often been likened to the twin strands of DNA for the American character. Jefferson believed in a decentralized government to guarantee liberty, while Hamilton believed in a strong central government to prevent chaos. Jefferson didn't care about criticism. Hamilton had a hair-trigger temper regarding criticism, which explains why he died in a duel to defend his honor with Aaron Burr, Jefferson's vice-president, in 1804.

I find myself identifying more with Jefferson, seeing the need for multiple centers of inquiry, loving vigorous debate, and not caring if somebody criticizes me or my ideas. Frank is more like Hamilton, believing science needs to speak with a unified voice, and bristling with indignation if he feels he's been unfairly criticized.

Maybe it's because I'm a woman who feels she's always been dismissed by the good ol' boys of science, but most of them don't impress me very much. For example, Frank will care about what John Coffin thinks about him, even though Frank's been proving Coffin wrong for nearly forty years, in things as important as whether human retroviruses exist. (Hello, HIV-AIDS, killer of more than thirty-five million people!) I look at John Coffin and see an arrogant misogynist.

Even though Jefferson and Hamilton were often on different sides of an issue, Jefferson respected Hamilton. At Monticello, Jefferson placed busts of himself and Hamilton staring at each other, as if realizing their two points of view would constitute the essential dialogue of this new country for centuries to come. Jefferson would later speak of Hamilton as a "singular character" who was possessed of "acute understanding" and was "disinterested, honest, and honorable." I could say all of this about Frank and more.

It's probably not a surprise that *Hamilton* is Frank's favorite musical and he's fond of quoting the lyrics, "Who lives? Who dies? Who tells your story?"

As I've said, I'm more of a Jeffersonian, and it's not simply because I did my undergraduate work at the University of Virginia, founded by Jefferson. The three accomplishments Jefferson wanted to be remembered for and listed on his tombstone were: "Author of the Declaration of American Independence, Of the Statute of Virginia for religious freedom & Father of the University of Virginia." Independence, freedom, and the pursuit of knowledge. That about sums me up.

I think it's probably accurate to call me more of a revolutionary and Frank more of a conservative. But the reality is more complex than those labels. I may be a revolutionary, but I understand the need for stability. And while Frank may be more conservative, he understands the need for change. We may often start from different perspectives, but after a good debate, we can usually settle on a reasonable course of action.

But in this current dark age of science, both the revolutionary and the conservative are exiled. The revolutionary is shouted down for their new ideas, and when the conservative asks for the evidence to support an existing policy, they're told the issue has already been settled. Stop with the questions! In place of the revolutionaries and conservatives in science, we now have the liars, the mercenaries, and the cowards.

Science can weather honest disagreements between researchers of integrity and intelligence, but it cannot survive this current plague of corruption.

* * *

None of my former colleagues call to ask me to join in their research. No college or university will offer me a teaching position. Instead, I'm blessed to work and often bicker about these issues with my longtime collaborator, Frank, in a small consulting practice, doing what we've done for the past thirty-five years, namely, trying to understand the process of disease, and how to end the needless suffering of so many.

In Bernard Malamud's classic baseball novel, *The Natural*, the hero is told, "We have two lives, Roy, the life we learn with and the life we live after that." One might say the life I learned with is the story Kent and I told in *PLAGUE*. This book you are now reading is about the life I lived after that, when we discovered corruption in many areas of science was widespread but realized there was also great reason for hope.

What seems like an ending is almost always a new beginning in some way.

I entered professional science on June 10, 1980, as a protein chemist working for the National Cancer Institute to purify Interferon, at the time a revolutionary cancer treatment. My eventual mentor at the National Cancer Institute, Frank, had been part of the team that discovered the first human retrovirus, HTLV-1 (human T-cell leukemia virus). We were well poised to fight the HIV-AIDS epidemic, which was looming in the future. I remember working at the National Cancer Institute in the mid-1980s and walking through crowds of angry AIDS activists who were shouting we weren't doing enough to cure their disease.

In 1991, when I defended my doctoral thesis, it was about how HIV hides from the immune system like a Trojan horse and how targeting drugs to this Trojan horse could turn this fatal disease into a manageable one. One week before my doctoral defense, the professional basketball player Magic Johnson tested positive for the HIV virus.

I was asked by my thesis committee members whether Magic Johnson would die of AIDS. My lengthy molecular biology response was simply that not only would he not die of AIDS, he would never develop AIDS, given that his infection was recently acquired, and the new drugs would silence the activity of the virus in order to avoid damage to his immune system. This was totally against the dogma of the time that these drugs were too dangerous and should only be given in the later stages of the disease. By that time, I argued, Magic would have no immune system with which to respond to the drugs.

More than twenty-five years later, Magic has not developed AIDS and lives well. As have millions of others who would have otherwise died. And we are doing even more. We have learned not only how to silence the virus but are discovering how to flush it out of its hiding places and eradicate it, so there can be an actual cure.

In an optimal setting, that's what science does.

It tells the truth and finds answers.

Even when that truth is a dark one, we must find a way to bring it into the light.

I've been asked why I'm writing this book. After all, hasn't my tale already been told? Our work was celebrated for a moment, then I was eliminated. End of story.

But just because science isn't paying attention to Frank and I doesn't mean we haven't been paying attention to the science. We understand so much better the inflammatory storm raging in the bodies of millions and how we might use things like cannabis, suramin, energy therapies, diet, and other natural products to quell this tempest.

We can put the ghosts of the past to rest and move forward into an unimaginably bright future of health for all.

CHAPTER ONE

A Scientist at Sea

It was late October of 2011, halfway between getting fired from the neuroimmune institute I cofounded and finding myself in jail, when I was riding my bike down Harbor Boulevard through the sand dunes of McGrath State Park in Oxnard, California.

Think of every fantasy you have of Southern California—the blue Pacific Ocean with whitecaps, a late fall breeze, the beach, parks where parents with their children were flying kites—and you have a pretty good picture of why I liked to take this route. That day I was biking from our boat-dock home located on a small canal to the PBYC where I was part of a group planning an annual sailing competition to benefit Caregivers, an organization that helps the elderly remain in their homes.

And what did I look like as I rode through some of the less-traveled areas near McGrath State Park? I was in my midfifties, stood five foot four, weighed around one hundred and forty pounds. I imagined I was probably indistinguishable from a great number of people as I pedaled along on my blue bike wearing an orange helmet and bright biking clothes.

Although I'd recently lost my job and was in the middle of a heated scientific controversy, I wasn't unduly worried. I was the principal investigator on federal grants worth approximately 1.5 to 2 million dollars a year for any university that hired me. I had interviews lined up at the University of California, Los Angeles (UCLA), the University of California, Santa Barbara, California State University, Channel Islands, as well as an opportunity in New York City at Mount Sinai Medical Center with Dr. Derek Enlander. We owned three homes, several cars, and a boat, had money in the bank, and my husband had a generous pension from the years he worked as the human resources manager for a major hospital.

The research institute I'd cofounded was housed at the University of Nevada, Reno. The man who'd hired me, Harvey Whittemore, was generally considered the most powerful person in the state. He would eventually spend eighteen months in federal prison for campaign finance violations involving Senator Harry Reid, at that time the majority leader of the United States Senate. These people who had once been my close friends had betrayed me and millions of others. I had refused their directive to be part of what I considered to be unethical and illegal actions. And I had not gone quietly into that good night. I had raged against the dying light of hope that had been so briefly kindled by our work, on behalf of a forgotten group of terribly sick people.

A white pickup truck with Nevada plates pulled ahead of me, parking in the bike lane. As I pedaled by the parked vehicle, I saw the driver holding up his cell phone, as if he were taking pictures of me. He was a large man with a beard, brown hair underneath a baseball cap, sunglasses, a tan, and he gave off an unmistakably creepy vibe. I couldn't help but notice he had a rifle mounted in his back window. This dance of following behind me, pulling ahead and parking to let me pass, then pulling out again happened several times before I crossed the street and rode opposite traffic and he drove off.

When I made it to the yacht club, I told the story to a friend. "It was really pretty strange," I said. "He just kept following me."

"You fool," said my friend. "You could disappear. All he has to do is make sure it's you, grab you, throw your bike in the dunes, throw your cell phone in the water, then when they find your body someday, people will say you killed yourself because things weren't working out in the XMRV study. So help me, God, if you ever ride that bike again, I will personally kill you. I'm driving you home. And from now on, you are never in a place alone where people like that can find you."

She was adamant, and I complied, realizing that one of my blind spots is not being able to see when people intend me harm. I was often referred to as a "lab rat," the designation given to those scientists who prefer to spend their time at the bench doing experiments rather than glad-handing politicians and donors or haranguing graduate students on the work for which the senior scientist will take credit. I preferred to be in the lab, hands-on, shoulder to shoulder with Frank, research assistants, and students, guiding them and challenging them, as they do the same, making sure the explanations I gave them and the conclusions we made were sound.

This is where I have spent most of my professional life with Frank, challenging the reigning dogma when the lens of a microscope tells a different story.

However, I was about to get an education in the dark arts of humanity, the landscape of fear and lies. I did not fully appreciate the power of those who wield these skills. I am not sure I have found my way back to the light.

I think more of us are under this spell than we realize.

* * *

How do you commit the perfect crime in science?

We're handicapped from the start because it's a question we never ask. For more than thirty years, Frank taught me and many others to record our data accurately, compare them with collaborators around the world, discard the outliers, and come to a consensus. We understand there are variations, but if the bulk of the evidence goes in a certain direction, we are confident we have a better understanding of human biological processes.

If only that were what happened in the real world.

In the real world there are corporations, be they pharmaceutical, agricultural, petroleum, or chemical companies, that have billions of dollars at stake in the work of scientists. If one has billions of dollars, he can use the dark arts of persuasion to hire public relations firms to tout your products, sow the seeds of doubt about those who question your products, buy advertising on news networks so they don't publicize negative stories unless they have no other choice, and donate to politicians of all ideologies. Then, once those politicians have been elected, they can write laws for the benefit of their generous donors. As it was put so eloquently in the seventeenth century by a prominent member of Queen Elizabeth's court, "If it prospers, none dare call it treason."

Against this financial and corporate juggernaut is the naive and inquisitive scientist. We are not taught to be fierce. We take no graduate courses in courage. We are encouraged to believe the raw data as long as all the experimental controls are used, and we report ALL of the data, even if we do not understand them.

I've often thought we in science would do well to follow the example of lawyers. In my conversations with attorneys, it's clear they relish intellectual combat. They will stand and defend the most hated individual in society, because they believe that person is genuinely innocent, or that a certain process must be followed before we can pronounce judgment. Frank taught me a love of such intellectual combat. In Frank's eyes, if you'd followed the scientific method, you had a duty to fiercely defend those data. And with Frank, you checked, rechecked, and triple-checked your data before he'd let you show them.

Once a colleague told us, "the most important data in a scientific paper is those data you don't show." That statement enraged Frank. He'd often say, "the best papers leave the readers asking more questions than they answer." We left in all the data in our October 8, 2009, *Science* paper, even those parts we did not understand at the time. Even though that paper ended my career, it speaks the truth to this day.

Those in the legal profession are taught to be fierce. I am thankful that Frank taught me to be as fierce as any lawyer.

The best scientists in history have been those who have similarly gone against the grain of traditional thought. Think of Galileo claiming the Sun did not revolve around the Earth, or Darwin challenging the Biblical idea that all creation, plants and animals, land and sea, were created in six days, and God rested on the seventh.

One day, as I was whining about the negative papers being published supposedly refuting the XMRV association with ME/CFS, Frank took me into his office and pointed to a file cabinet in the corner. He opened the drawers containing published papers saying he was wrong about T-cell growth factor (interleukin-2), or HTLV-1, causing Adult T-Cell Leukemia. One paper was published that very month! He said, "If you can't stand the heat, get out of the kitchen. Now, let's get back to work."

I encourage you to consider the scientific questions raised in this book the same way you would follow one of the high-profile criminal cases that from time to time consume so much of our national attention. You understand a claim will be made. This person is accused of killing another person. You will hear the evidence presented, watch it be challenged by the other side, then come to your own determination as to which pieces of evidence are credible and which are not. It is a methodical process. After each side has introduced their evidence and dealt with the challenges to its credibility by the opposing side, you arrive at your conclusion.

Let me make the claim that underlies everything that follows.

Science is being corrupted by the influence of corporate money. This corruption is leading directly to our poor health, whether it be the epidemic of obesity; neurological diseases like autism, Alzheimer's, Parkinson's, and multiple sclerosis; the explosion of cancers; or mental problems among the young, including school shooters. There are some who claim this is leading to a culling, if not the mass extinction, of humanity.

Based upon all we have experienced, I find it difficult to counter this troubling narrative.

I entered all of this as naive as a first-year graduate student.

I did not believe science to be as fundamentally corrupt regarding our health as I have now come to believe. I want you to consider me as the young boy in Hans Christian Anderson's tale, *The Emperor's New Clothes*. In the story, the Emperor is told by the crooks that they were creating a set of clothes for him that could be seen by only the finest people. All the people around the Emperor claimed they saw the beautiful clothes because they wanted everybody to believe they were the finest people. Only a young boy, unconcerned by what others thought of him, pointed out that the Emperor was naked.

If you are continuing to read this book, you are effectively impaneled as the jury on a claim that science has chosen a misguided path. You've taken an implicit pledge to listen to what we and others say with an open mind. We did not come to our conclusions easily.

I do not expect you to, either.

But let us begin this journey together.

I did not know if I could write this book. Many of the incidents are so troubling to recount that I worry about suffering from post-traumatic stress disorder, the condition so often found in soldiers, police, or firefighters, who serve on the very front lines of conflict. This is the story of a battle being fought by a few brave scientists against an enemy with almost limitless resources.

Science may be agnostic in this struggle, but I am not.

I'm a person of faith and believe God wants humanity to enjoy good health, not suffer.

Sometimes people ask me how I'm still alive, and I reply, "God has a sense of humor." I do not know my ultimate fate or how I will be judged by this world. It doesn't matter.

However, I know one day I will stand before God, who will ask if I was obedient and served as He requires. What I share in the pages that follow is the account I would give the Almighty on Judgment Day.

* * *

The pounding on the door of our boat-dock home located on Jamestown Way in Oxnard began at about 5:00 a.m. on November 9, 2011.

I was in the shower, and my husband, David, waking and finding me not next to him, figured I was already on my way to work, as I was most mornings. I like to get an early start.

Always have. Even though David wears hearing aids, he does not wear them at night, and he stumbled out of bed to walk downstairs, not realizing I was in the bathroom.

A man was at the door, wearing a badge and claiming he had a legal document to serve to Judy Mikovits.

"She's not here," David replied, wearily, wearing only his boxers and a t-shirt. "She's long gone. She'll be back here at about eight o'clock. You're welcome to come back tomorrow or wait."

The man declined the offer to return and waited out in front of his car.

That morning I was supposed to have a meeting at UCLA, accompanied by my good friend, Ken, with whom I had previously worked at EpiGenX Pharmaceuticals in Santa Barbara. UCLA was sixty miles away, and that distance in Los Angeles rush hour traffic is not a treat. There was also the possibility that Ken and I would meet up later in the day with Patrick Soon-Shiong, the Chinese billionaire who would eventually purchase the *Los Angeles Times*, to discuss a potential job working for one of his companies. Prior to amassing great wealth, Soon-Shiong had been a transplant surgeon and founded a successful biotechnology company. Ken thought the three of us would speak a similar language.

David walked up the stairs as I came out of the bathroom, just about ready to leave. I asked, "What was that about?"

David nearly jumped out of his skin. I'd like to say these things happen because my husband is twenty years older than I, but I've known enough couples to realize this is a relatively common experience for many.

After he calmed down, he explained to me what had happened.

"That's odd," I said, recalling I'd been threatened with a lawsuit by my former employers on November 2. The letter had given me only forty-eight hours to respond, and I'd complied, using my friend Lois, an attorney who suffers from ME/CFS. We'd faxed the reply on November 4 from my friend Lilly's house, well within the deadline.

After the incident with the creepy man in the white pickup truck with Nevada plates, I started to become suspicious of other events. Our boat-dock home was an end unit in a row with other houses. Just across from us was a small greenbelt and next to it, another end-unit home that had been unoccupied for a long time. In October, it was suddenly occupied, and the new residents installed bright lights that appeared to be focused on our house. Residents often put such lights on their houses to illuminate the water of the harbor, but these seemed oddly angled. I've always enjoyed natural sunlight, so I don't have any shades or curtains on my windows, so it was as if I were living under a spotlight.

Mike Hugo, my attorney, was later able to get through discovery an admission that during this time I was under surveillance by police from Nevada and California and local law enforcement.

I quickly gave a call to Ken, telling him what had happened and explaining I probably couldn't make the interview with UCLA that day.

Ken was immediately on high alert. If anybody understood the high stakes of where my research was leading, it was Ken. He was a money guy who knew our discovery was worth billions, and it was clear to him what this cover-up was about. The intellectual property! Not only had we discovered a new retrovirus family, but our colleagues were saying it had probably been spread through the population via vaccinations and had likely infected more than ten million Americans.

Could I be in any more trouble?

"Get rid of your cell phone," he said. "Take the battery out of it and throw it in the water. Do not use that phone again. They can track you with it."

Neither Ken nor I was a secret agent. I was a scientist, and he was a money guy with a background in public health.

"Okay," I answered.

"I'll make an excuse for you. But you need to get out now."

We talked for a few more seconds, then I quickly hung up and took the battery out of the phone.

My mind immediately flashed to a padded 5 x 7 manila envelope that a ME/CFS patient and autism mother from one of the family studies had sent me in the summer of 2011.

We'd been focused on the possible connection between ME/CFS and autism, but the truth is our work was casting a much broader net.

Infection by a retrovirus can cause a myriad of diseases, depending upon each person's unique genetic vulnerabilities.

The bulk of my twenty years spent in government science had been at the National Cancer Institute. It was a pattern of unusual cancers among long-term sufferers of ME/CFS that first caused me to become interested in the condition. As with HIV-AIDS, mothers with XMRV could pass the virus directly to their children. Transmission to the spouses, although possible, was less likely.

In the padded 5 x 7 manila envelope, the mother had sent me several one hundred-dollar bills, a portable camping potty, a fake pen with a recording device and camera, a go-phone with minutes on a card, and a note that read, "You really don't understand, but you're going to need these."

At the time I'd received it, I'd fretted with David about it. "We have plenty of money in the bank. Why do I need ten one hundred-dollar bills? We must send back this money to that sweet mom." I'd called up the mother and offered to send it back, but she refused. My husband often needs to use the bathroom on long trips, so I thought I'd at least keep the portable camping potty. Practical Judy.

I put the card in the go-phone, got it working, then called Frank, who I knew would be at his desk in Frederick, MD, early in the morning. Before he was forced to retire in 2013, Frank would spend a total of thirty-nine years at the National Cancer Institute. I quickly explained what had taken place and asked his advice.

"You dummy," he said. "You've got a boat and you live on the water. They can't take somebody off the water. You can escape from your house."

It was a great idea, and I quickly put a plan into action.

My stepdaughter, Elizabeth, was staying in our second bedroom at the time.

Depending on the time of year, she's either six or seven years younger than I, and we have a similar build and hair color. It was her birthday, and we were planning to take her to lunch that day. David went into her room and roused her and asked her to come downstairs.

I laid out the plan to David and Elizabeth. "The two of you will leave the house and go for a walk in the neighborhood. Let's figure out what's going on."

"I don't want to go for a walk!" David complained.

"I don't want to go, either," Elizabeth joined in.

"Look, it's going to be fine. They're just trying to serve me with something. Let's figure out what's going on."

The two of them got ready and left the house. After getting a short distance away, three men approached them, and one of them was the same who had knocked on our door. "Judy Mikovits, we are serving you with a lawsuit," he said, producing a piece of paper.

"I'm not Judy Mikovits," said Elizabeth, pulling out her driver's license and showing it to them. "He's my daddy and it's my birthday," she said with a laugh, throwing them off their game. After briefly examining her driver's license, they let them go.

When they returned to the house, it was clear to me that I was surrounded. The water was the only way out. "Elizabeth, I want you to go out on the deck. Let them see you. David, I want you to get the Baby Jonah ready for a boat ride."

"I don't want to go for a boat ride!" David protested.

"You're going to take your daughter for a boat ride because it's her birthday and we promised to take her for lunch." He complied, and I waited inside, taking care to stay away from any of the windows where I could be observed. I went upstairs and packed a backpack. By this time, it was about eleven in the morning. (Noon would be low tide.)

When David came back saying the boat was ready, I asked him what Elizabeth was wearing.

"I don't know," he replied.

This was killing me. In light of what I'd experienced over the past few weeks, I'd imagined the situation would be clearer to them.

David left and returned, saying she was wearing black yoga pants and a dark t-shirt. I found some clothes that approximately matched those, along with a dark blue t-shirt given to me by one of the patients that read, "CSI: Can't Stand Idiots." Nothing like a little humor when you're surrounded, right?

I had two identical baseball caps bearing the sail number of my friend's boat and gave one to David for Elizabeth to wear, since it was windy that day. "Here's what I want you to do. Give her the cap, go to the boat for a few minutes, and start the engine. Call out to Elizabeth that you want to take her to lunch. She'll say it's cold and she doesn't want to go. You just tell her to go in and get a jacket and come back. That's when we'll make the switch."

David seemed to get the idea, though he thought it silly. Within five minutes, all of this had taken place and Elizabeth walked into the house. I looked at what she was wearing, the jacket I'd put on, and we were a close match. I waited for a few moments, then, clutching the backpack, walked out the back door to the dock. I jumped on the boat, David cast off the line, and we started moving down the channel, hugging the left seawall.

When we got to the main channel, David floored the accelerator. He loved to go fast on that thirteen-foot Boston Whaler. He looked at me and, in the fake Russian accent he often likes to adopt, said, "Katarina, vee have escaped! But vee don't know where to go. What shall vee do?"

"I've got an idea."

I called my dear friend Robin, who owns a thirty-eight-foot sailboat docked in the Channel Islands Harbor. The boat sleeps five people and David and I had been sailing on it often with Robin and her husband, Steve.

I told her we had an issue and asked if I could stay on it for a few days. She said sure, and I then asked if she could bring some food.

"And one other thing. Is there vodka on board?"

"There's always vodka on board," she replied.

Even though it was still morning, I knew I'd have a tough time sleeping that night.

David knew the route to the Channel Islands Harbor (located on the coast of California) and Robin's boat. Within about fifteen minutes, he'd pulled up to her boat. I gave him a few of the hundred-dollar bills from my envelope so he wouldn't need to use a credit card. At this time, I figured I was simply dealing with some Harvey Whittemore goons from the University of Nevada, Reno, and a couple local rent-a-cops determined to intimidate me. We simply needed to find an attorney who could practice in both California and Nevada. Then I would be safe. I told David he needed to avoid Robin's boat until he had found an attorney.

After getting onto the boat, I headed down into one of the cabins and pulled out my go-phone. I called ME/CFS patient and friend, Jeanette. She and her husband, Ed, are both attorneys in San Francisco. After I explained what had transpired, she said she'd start looking for a local attorney who could handle my case.

When I finished my call with Jeannette, I paused for a moment to catch my breath. It was a very windy day, the rigging on the mast was clanging, the boat was slowly rocking, and I was completely terrified. This is not how a PhD scientist, with more than fifty peer-reviewed publications to her name, expects to find herself, especially when she recently led the team that produced a ground-breaking publication in the world's most prestigious journal of original research.

Fighting down my rising sense of panic, I called Frank to let him know I was safe. He was happy to hear the news but had another pressing issue on his mind. The journal *Science* had called and was pressing him to retract our entire article published in October 2009, which showed isolation of a new family of human retroviruses and an association with ME/CFS. We had known this fight was coming, for many of the reasons I have previously mentioned.

If our data were allowed to stand, the inevitable result would be a massive financial catastrophe for the world's pharmaceutical companies because of their negligent use of animal cell cultures to produce vaccines and other pharmaceutical products. As any criminal lawyer will tell you, the fight is always about motive and opportunity.

I have just given you the motive.

Let me tell you about the unique opportunity created by XMRV.

In March of 2006, Dr. Robert Silverman of the highly regarded Cleveland Clinic and his team published detection of nucleic acid sequences related to murine leukemia viruses in samples of tissues from men who had prostate cancer. They named the virus XMRV. The idea that viruses may predispose to cancer has been widely upheld and is one of the most active areas of research in medicine.

There was only one problem with Silverman's report of a new human retrovirus causing aggressive prostate cancer. They never isolated the virus and showed it to be infectious and transmissible. They had only detected a few hundred base-pairs of XMRV, and from that they cloned the remaining eight thousand base pairs in the laboratory using only those few hundred base pairs of the DNA they obtained from the biopsies.

Because the tissue biopsies contained only a tiny amount of DNA, Silverman and his colleagues had to piece together the virus from several biopsies from different patients in order to get enough DNA to put together a complete sequence and generate an infectious molecular clone.

An additional level of complexity is what we consider genetic stability.

For example, between humans and chimpanzees, there is a 99 percent similarity in genes, but they are dramatically different creatures. However, the reverse transcriptase enzyme of retroviruses is error-prone. This causes wide genetic difference in the virus.

Retroviruses with up to 10 percent variance in their genetic codes will still be classified in the same retrovirus family, perhaps simply a different strain or clade, just as with the HIV or HTLV families of viruses.

What Silverman's colleagues did was to take DNA from three different patient biopsies and splice them together in a Frankenstein fashion to create an infectious molecular clone that he referred to as XMRV. In fact, we were the first to isolate the natural XMRV from humans, as we reported in *Science*. Silverman's stitched-together Frankenstein monster clone XMRV, which had never existed in nature, was given the designation VP62.

Silverman did not tell us he created that infectious molecular clone from three different biopsies until June of 2011. He only admitted his mea culpa to Frank when the inescapable conclusion of all the data we did not understand, but staunchly refused to take out of the *Science* paper, eventually revealed his deceit.

I consider Silverman's failure to tell the world what he did in 2006 a crime of the greatest magnitude and should have resulted in his being driven out of science.

It was not a mistake.

He concealed a material fact.

VP62 had dramatically different sequence and growth characteristics than our naturally occurring XMRV isolates, which we obtained from ME/CFS patients, cancer patients, and children with autism. The deceit was clear, as the serology test for VP62 developed by Abbott, which funded millions to the Silverman lab, never detected a single positive. Yet, the serology test we published in *Science* detected most of the variant strains.

Another example was that VP62 replicated at a rate at least twenty-five times faster than natural isolates of XMRV, many of which were defective, as with most retroviruses, including HIV. That meant that if VP62 was anywhere near a representative of a natural XMRV infection, it would soon overrun it like an invasive species.

In 2011, we discovered that both VP62 and XMRV had the capacity to become aerosolized, meaning they could simply drift through the air, going wherever the slightest breeze might blow them. The guardians of corrupt science realized they had unleashed VP62 like an assassin and it could destroy evidence of natural XMRV if the two samples were housed in the same facility.

My greatest concern was that well-meaning efforts to create weakened viruses for vaccines had created new problems. Had science properly considered the question of whether any mixing of human and animal tissue brought with it the inevitable risk of transferring animal viruses to humans? Or that these animal viruses would combine with human viruses to create new pathogens?

Frank was keenly aware of that simple fact, as through the summer of 2011 we had been working feverishly going over our samples to show they contained XMRVs and not Silverman's VP62 molecular clone. We proved that repeatedly. I'd had a scientific debate with staunch supporter turned fiercest critic, John Coffin, on September 22, 2011, in Ottawa, Canada, on the issue and had come away the clear victor. In my conclusion, I asked him, "How many XMRVs have we created, John? How many?" This debate in a session of the IACFS biannual meeting was supposed to be about whether the scientific community had a diagnostic screening test for XMRV in the blood supply. Why had *Science* rushed this fraudulent paper that targeted the patient community? It was NOT an association study. There were only fifteen patients. And besides, how can you have an association study with a Frankenstein virus constructed in a lab that had nothing to do with the natural strain of XMRVs we isolated?

If these fears were true, was science standing up and admitting the mistake? My worry was that researchers like Coffin didn't want to rock the boat. They didn't want to be the ones to stand up and say science had made a terrible mistake and possibly injured millions of people.

This injustice surely could not be allowed to stand.

Who was I, a middle-aged female scientist at an upstart institute out of Nevada, to tell the old boys network of science that they and their predecessors had made a terrible mistake?

And I was demanding it had to be stopped. I refused to retract the paper, for the data proved all I had said, and to this day it remains right and true.

By Sunday, November 13, David found an attorney, and I felt safe enough to come out of hiding, return to my home, and go to church services that afternoon.

The following Friday afternoon as we returned from a beach walk, three police cars descended on my house. I was arrested, taken away to jail, and held without bail for five days. The police searched my house from top to bottom, leaving papers strewn everywhere, claiming I had taken notebooks that were obligation to retain as the principal investigator on two government grants.

They did not find the notebooks during that search.

That same day, police raided the home of my friend Lilly and forced her to sit in a chair for several hours while they conducted a search of her house. The only explanation I have for this action is that I faxed my response to their civil claim from her house. No notebooks were found. They were not there. Never were. I left them in my office the day of September 29, 2011. I believe the Whittemores knew it and fabricated the events of that day in order to cover their crimes. The crimes as detailed in my scientific debate with John Coffin were memorialized in an article published by *Science* on September 30, 2011, the day after I was fired. The article, written by Jon Cohen, ended with "she hopes to have full sequences of her new viruses in a couple of weeks."[1] That was why I needed to be fired, and when I wouldn't stop my investigations, they carried out the threat to destroy my career.

During my time in jail, the people who engineered my arrest and harassment, the powerful Whittemore family of Nevada, had David running all over the place trying to release the hold on my bail. Despite a thorough police search of my residence, the notebooks in question suddenly appeared at my house on Monday evening, placed in a bag from my residence in Reno when I was fired in late September of 2011. Now I think I fully appreciate the desperation of my former employers, the Whittemores. Fire me on the

29ᵗʰ. Immediately lock down my lab and two offices. I had shown Harvey Whittemore what I believed to be the accurate sequences of XMRV in late August of 2011. Frank had that information, as well. Preventing me from accessing my notebooks, as has been done since I was fired, keeps me from having the information I need to defend myself.

There are only two possibilities.

The first possibility is I'm lying, and the police were incompetent when they searched my house during my arrest.

The second is I'm telling the truth.

Here are some facts I want you to consider.

Charges were never brought against me in any court nor was any trial ever held. I'm not like one of those mobsters who say they've never been convicted, but they've sat through several trials where government lawyers did their best to convict the defendant. In fact, no prosecutor ever looked at the facts of the case at all. David was called by Harvey Whittemore and told if I did not find the notebooks I would remain in that jail throughout the Thanksgiving holiday. When David found them neatly packed in that linen beach bag and frantically took them to the jail well after midnight, no one had any idea what he was talking about. They sent him back to the arresting officers.

This is how the chain of events were described in the False Claims/ RICO law suit I filed against Harvey Whittemore and others on July 27, 2015 (originally on November 17, 2014):

> On November 21, 2011, the plaintiff's husband received a phone call from a representative of HW [Harvey Whittemore], AW [Annette Whittemore], Kinne [Carli West Kinne], Lombardi [Vince Lombardi] and Hillerby [Mike Hillerby], to discuss the fact that plaintiff would likely remain in jail through the Thanksgiving holiday, which was in two days, unless he returned the notebooks.
>
> Having nearly completed the entire task of reorganizing all the materials, clothing, books, papers, and other possessions that had been strewn about the house by the UNRPD [University of Reno, Police Department] officers in the warrantless and illegal search, the plaintiff's husband assured the representative of HW, AW, Kinne, Lombardi, and Hillerby, that he had been through the entire house and that the notebooks were not there. He assured the representative that if the plaintiff had the notebooks, neither she nor he were aware of it, and that they were not in the house.
>
> At that time, the representative of HW, AW, Kinne, Lombardi, and Hillerby told the plaintiff's husband, "David, listen very closely to me. You

DO have them. I am telling you. Now go and find them and return them to get Judy out of jail.

The men hung up the phone and the plaintiff's husband sat in complete perplexity at the entire conversation, knowing he had scoured the entire house as he replaced items in drawers, closets, shelves and table tops.

The following morning, the plaintiff's husband awoke and reinitiated his search, looking for places that the plaintiff may have secreted the notebooks, all the while replaying the conversation with the representative of HW, AW, Kinne, Lombardi, and Hillerby, in his mind.

As the plaintiff's husband began to look through cabinets, book shelves and drawers for the notebooks that the representative of HW, AW, Lombardi and Hillerby insisted were in the house, he came up empty. Repeatedly doubting his sanity as he continued the same search that he and the police had each previously conducted, somehow expecting or hoping for a different outcome, he was rapidly becoming disheartened as he began to dread— Thanksgiving—which he knew would be the loneliest day of his life.

While searching through one of the guest room closets, the plaintiff's husband discovered a canvas beach bag with JAM embroidered on the side, that he had not seen previously, and that was not inventoried as part of the search. Even more suspicious was the fact that the bag was sitting in the front and center of the closet as if it were the last item placed therein. Inside the bag were the plaintiff's notebooks.

The notebooks were planted in the closet by the representative of HW, AW, Kinne, Lombardi and Hillerby, or by other agents of HW, AW, Lombardi and Hillerby.[2]

I did not have those notebooks, and I did not have that canvas beach bag. Those notebooks were secured on September 29, 2011, by my research associate, Max Pfost, who suspected the Whittemores would try to sell a non-validated test and that canvas beach bag was in my place in Reno, Nevada. I believe Max was forced to give those notebooks to Harvey Whittemore, who then decided to use them in an attempt to frame me by claiming I'd stolen them and planting them in my house while I was in jail and my husband was racing all over the place trying to find a lawyer. Harvey or somebody associated with him must have gone into my place in Reno and taken that bag so they could put the notebooks inside and plant them in that closet.

I appeared before the Ventura County judge late Tuesday afternoon, when Jon Cohen of *Science* was waiting, petitioning the court for a picture of me in my orange jumpsuit with my hands and feet shackled. Thankfully, the judge refused. But that would not deter Jon Cohen from sending the message to all scientists of their fate if they went down the same ME/CFS retrovirus road.

I was forced to return to Nevada under the threat of arrest, where a mug shot was taken and published in a scandalous article in *Science* by Jon Cohen, designed to discredit me in the eyes of the scientific community and the greater world beyond. *Science* had achieved its goal and had the ammunition to force the retraction of the paper. My scientific career was destroyed.

My life savings in the bank?

Gone.

Our homes?

Gone.

We spent it on lawyers trying to bring cases for the violation of my civil rights, the false claims, and to end this plague that reaches into every town and city in this country.

Despite spending all that money without obtaining justice, I was forced to declare bankruptcy. Not because I didn't have the money. My attorneys believed that if I appeared, "new" evidence would be found against me and I would end up in jail in Nevada, and possibly die under mysterious circumstances.

"I don't need to file bankruptcy," I said. "I have a perfect credit score. I'll sell my houses and take my ninety-seven witnesses to Reno to the damages hearing of that fraudulent case. I'll not only prove the science was right, but that crimes were committed against this patient population."

Dennis Jones, my civil attorney, was firm. "Let me tell you what will happen if you do that. When you step on the Reno courthouse steps, you will be immediately arrested by the district attorney, who will claim they have new evidence against you."

"That's ridiculous," I replied. "There's no new evidence."

Dennis leaned forward and coldly said, "There was no evidence the first time, was there?"

Tears welled up in my eyes. It was the only time I'd cried during the whole ordeal. I knew it almost killed my husband the first time I'd been jailed. This time, it certainly would.

I declared bankruptcy. I believe I saved the life of my husband by this action. It was a strategic loss. I did not want to do it. However, sometimes you must lose a battle to win a war.

I have continued to fight.

Have I spent many hours with FBI agents telling them about the corruption of science that I believe existed at the University of Nevada, Reno, NIH, CDC, FDA, and among our National Academy of Science leaders? Yes, I have. In fall 2018, I was notified that the false claims whistleblower case, which had been under seal for three years, was no longer under seal, and I could serve the defendants.

I filed a motion for clarification. Which defendants do I serve? On April 11, 2019, I filed a criminal affidavit for extortion and obstruction of justice.

As of this writing, I have not received an answer to either. I doubt justice will ever be served on this Earth. The United States has abandoned any allegiance to either Jeffersonian or Hamiltonian principles. I often think of Clint Eastwood's line in the movie *Unforgiven* as the sheriff begs for his life, saying he doesn't deserve this. Eastwood responds, "deserving's got nothing to do with it."

Kent tells me that the freedom to publish a book and lay out your side of the story may be the last actual freedom we have left in this country. Maybe he's right. The courts are corrupt, the media, politicians, scientists, and physicians are bought off or bullied into silence.

It's almost enough to turn an honest scientist into an outlaw.

CHAPTER TWO

A Rebel from the Start

I was in the office of my boss, Russ, for one of our regular meetings. As I stood in front of his desk, clutching my lab book, he said something I've never forgotten. "You have a moral, legal, and ethical responsibility to do exactly as I tell you."

It was the summer of 1987, I was twenty-nine years old, and I was at Upjohn Pharmaceuticals, located in Kalamazoo, Michigan. The previous year I'd taken a job at Upjohn as a lab technician in their Quality Control Department after leaving my job at Fort Detrick, Maryland, as the Biological Response Modifiers had been dissolved. I didn't know it when Russ made that statement, but I was already nearing the end of my career at Upjohn.

Where is the good place in science?

I asked that question many times during my decades in research. Was it in government-sponsored science, which is supposed to be free of bias or politics? Or was it in industry, supported by the profit motive, which puts serious money behind advances that have the potential to change people's lives? I have come to believe that both government-sponsored science and industry have the potential for enormous good. But both can also easily go awry if those in charge lack integrity, which seems to be increasingly rare as one goes up the ladder in either government-sponsored science or industry.

I write this book in the hope that science can be guided back to its founding principles.

I wanted to be a doctor ever since I was a ten-year-old girl and watched helplessly as my beloved grandfather wasted away from lung cancer. I used to listen to baseball games with him, and he kindled my lifelong love of sports. When I was a senior at the University of Virginia in 1980, I read a *Time* magazine cover story on interferon, which at the time was being promoted as a potential breakthrough in cancer treatment. I got a job later that year as a lab technician at the National Cancer Institute in Frederick, Maryland, and my career in science was launched. And what was my job? Purifying interferon-alpha.

The program I loved the most at the National Cancer Institute was the Biological Response Modifiers program, an interdisciplinary team of PhDs, MDs, nurse practitioners, and lab techs like me at Fort Detrick. During that time, we were working on adoptive transfers for AIDS patients, trying to understand how interferon-alpha might help, or whether the immune markers like IL-2 and other cytokines might provide the clues we needed in order to save their lives.

It was also a tumultuous time, as a controversy erupted over whether Dr. Robert Gallo, the most quoted scientist of the 1980s and 1990s, had tried to claim credit for the discovery of the HIV virus from French scientist Luc Montagnier.

Was it an attempted theft or just an honest mistake?

Montagnier would eventually win the Nobel Prize in 2008 for the discovery of the HIV virus, and Gallo's name would be conspicuously absent. I developed strong opinions about Gallo and what scientists like him were doing to the profession.

In 1986, the government in its infinite wisdom decided to eliminate the Biological Response Modifiers program. I would have to find a new position at the Institute. Also, around that time, I observed a senior researcher directing a young Japanese postdoc to change data in an experiment. The postdoc was clearly troubled by the order. Shortly after that exchange, he committed suicide by drinking sodium azide, a white solid that uncouples the electron transport chain, causing the person to suffocate and die. I went to the program director (Frank's boss) and told him what I had witnessed. "I know why he killed himself," I said. My boss was uninterested. The senior researcher got results. The young Japanese postdoc was written off as suffering from emotional problems, leaving behind a wife and two small children.

Later that day I got a call from one of my former coworkers, who was working for Upjohn. "Judy, we've got a job here for you and you'd really love it." I was born in Michigan and had grown up rooting for the Detroit Tigers. Michigan was home. And the money was a lot better than working for the government.

"I'm there," I said.

* * *

I was something of an oddball at Upjohn, but not for the reasons you might think.

During most of my time at the National Cancer Institute, I worked for Frank, and we had a natural sympathy for and common approach to science. Frank and I thought nothing of arriving at work around four or five in the morning and setting up our experiments, then working until six at night.

Science was what we loved to do.

Industry doesn't work like that. At Upjohn researchers came in at about eighty-thirty in the morning, worked until nine-forty-five, took a fifteen-minute break, went back and worked until a little before noon, then took off for a thirty-minute lunch, had an afternoon break around two, then went back to work for the rest of the day. In the summertime at Upjohn, we got off at three-thirty. The company had a number of intramural sports teams, and I played on softball teams, ice hockey in the winter, and soccer. I laugh when I say I played soccer, because I rarely ever touched the ball.

Our soccer coach was a tall, handsome black man named Wayne, who oversaw human resources. I used to run a lot and ride my bike, and I had quite a bit of endurance. I'd never played soccer, had little coordination, but I was tenacious. Wayne saw that in me and had me shadow the other team's best offensive players. I'd stick to them like superglue, and in frustration they'd call out, "Get off of me! Get off of me!"

We won games simply because I kept their best players from controlling the ball.

It was Wayne who first let me know my work ethic was causing a personnel problem. I'd come in three or four hours before my boss, Russ, arrived at nine. Nine in the morning was like the middle of the day to me. We were studying a recently marketed bovine growth hormone. Many other pharmaceutical companies were marketing a similar product. The claim was that the hormone increased milk production without any side effects on the cattle. I thought it was a great idea, but Upjohn didn't have a biological division to study the effects of its products on cell lines. They had a number of superb chemists and did excellent work with their High-Performance Liquid Chromatography system as well as their Mass Spectrometer, but they really didn't have researchers with a great deal of biological experience. That was one of the reasons I was so attractive to Upjohn. Part of what I was doing by coming in so early was designing the same type of biological assays I'd done at the National Cancer Institute.

"Why is it a problem?" I asked Wayne when he told me there was concern that I was coming in so early.

"People don't know what you're doing," he replied.

"I'm working."

"You need to let people know what you're doing," said Wayne. He thought about it for a moment. "How about you check in with Russ on a regular basis?"

"Mr. Personality?" Russ was a little guy, I don't think I ever saw him smile or laugh, and he didn't talk much. He was just a few years older than I and had recently received his PhD. He had a reddish beard and mustache, wore glasses, his lips often seemed clenched, and when you talked to him, he had a tendency to look away. Yes, sir, he was a real people person.

I went and asked Russ if he was okay with my coming in early to do work. He said it was fine. I also asked if we could meet a few times a week, so I could update him on my work. He agreed but showed no visible emotion.

However, another issue arose at Upjohn that drew me out of my work for Russ and delayed our inevitable confrontation over bovine growth hormone.

* * *

The product was called ATGAM, a drug for transplant recipients, and it was derived from human blood. But this was 1986 and the AIDS epidemic was in full swing, infecting hundreds of thousands of Americans every year, and creating a plague of slow death unlike anything seen in the United States for decades. HIV was a blood-borne pathogen, and so while it had been confirmed that the virus could pass from one individual to another through unprotected sex, the sharing of needles, as well as through blood transfusions, nobody knew if it would survive the manufacturing process for a product like ATGAM, produced from human blood.

The "official" government line was that HIV had not made it to Michigan and was confined to places like New York City and San Francisco.

To their credit, Upjohn did not share the same sentiment.

They were concerned that the possibility existed that the blood they were getting from Michigan residents to make their product might contain HIV. The hairdresser I started using when I arrived in Michigan told me he was HIV-positive and would eventually die from AIDS.

Since I had come from the National Cancer Institute, I had the kind of background that could help answer the question of whether the manufacturing process would decontaminate any HIV that did make its way into the final product. I quickly started interacting with a scientist named Bob, a wonderful man, who headed up a different division at Upjohn. He had a brilliant mind, was an opera singer, and just a delight to work with.

I told him that in order to determine if their product was safe, all we needed to do was spike their raw blood product with HIV samples we could obtain from the National Cancer Institute, then after each stage of the production process we would test to see if there had been at least a six-log reduction in the presence of the virus. He thought it was a great idea and quickly agreed.

I ran into trouble when I contacted the National Cancer Institute and asked if we could have some HIV samples sent to the Upjohn facility in Kalamazoo. Since HIV was not officially recognized as existing in Michigan, I could not bring any samples into the state. Did I want to be forever known as the woman who brought HIV to Michigan?

Bob came up with an ingenious, although unconventional solution. Upjohn had a Lear jet that they often used to shuttle their executives and scientists to Washington, DC, to meet with officials of the Food and Drug Administration to discuss their products.

Did I want to start hopping on the Lear jet so I could perform my experiments at the National Cancer Institute?

Frank would assist me in the effort, since I was still simply a lowly lab technician. (Poor Frank! Like a bad penny, I just kept coming back!) On Monday mornings I would often board the Upjohn Lear jet with a backpack over my shoulder, fly to Washington (Reagan) National Airport, where I would be met on the runway with my rental car, then drive to the National Cancer Institute in Frederick, Maryland, to perform the experiments or have meetings. Sometimes I'd fly back to Kalamazoo later that day, or sometimes I'd stay until the end of the week, flying back on Friday night. During those times when I was able to stay longer, I could visit socially with my former coworkers, or spend time with my mother and stepfather, Ken.

I also must comment on the somewhat clandestine nature of my flights to Washington.

Here I was, a twenty-eight-year-old kid, jumping on a Lear jet with the top executives of Upjohn, and not really being able to talk to my fellow colleagues about what I was doing. I usually dress casual, wear jeans, and often a baseball cap, and there I was in a small Lear jet with the top executives of the company. Even at a young age I was a pretty good conversationalist, and soon I was on friendly terms with a good number of the power players at Upjohn.

It was nice to be working with Frank again. In a few weeks, we had the experiments all set up and were in the midst of our investigation. I'm sure it helped that our testing revealed that their manufacturing process did a good job at removing the HIV virus from any potentially contaminated blood. I suggested a few minor changes, such as getting the temperature up for some of their processes to add an extra layer of security, but it was already safe.

Within my first few months at Upjohn, I'd already distinguished myself as a creative researcher, mixed with the top executives, made a lifelong friend in Bob (years later in 1992, Frank and I invited Bob to speak at a conference in Genoa, Italy, that Frank was chairing, as Upjohn had done some excellent work with HIV), and had a lot of fun playing on the intramural sports teams, all without having a PhD.

But it was time to get back to the job they hired me for, namely, working quality control, establishing a biological quality control division with cell lines on which to test their genetically modified biological products, investigating the claims for the safety of their bovine growth hormone.

* * *

At the time, I thought it was a great idea to use bovine growth hormone to increase milk production and help cattle grow to maturity quicker.

My job was to make sure it was safe.

The claim was that bovine growth hormone did not affect human cells. My job as quality control was to see if it was true or not.

I set up the experiments, using several different cell lines, added the bovine growth hormone, and waited to see what happened.

The claim that bovine growth hormone didn't affect the cells was false.

One of the first things I saw in the cell cultures was the bovine growth hormone was affecting the morphology (appearance) of the adipocytes, commonly known as the fat cells. Simply put, the fat cells were changing their appearance and not looking like healthy fat cells. In addition, when I tested to see whether the fat cells were producing the typical molecules of a healthy fat cell, impacting things like communication with other cells, I found there was a significant difference. The fat cells that had been treated with bovine growth hormone were producing different molecules. In all likelihood, these different molecules were affecting the function of other cell types.

I saw several examples of cells that had a condition called *blebbing*, in which a cell detaches its cytoskeleton from the cellular membrane, causing the membrane to swell into spherical bubbles. The proper working of the cellular membrane is critical in communicating with other cells, so what I was observing was probably causing a breakdown in cellular communication. I also observed some large neurite outgrowths, stringy-looking cells that were abnormal.

My research was leading me to a simple, but dangerous conclusion.

Bovine growth hormone profoundly affected human cell cultures.

A great deal of my work at the National Cancer Institute had centered on immune dysfunction caused by retroviral infection. I learned not only how to identify them, but also what conditions were likely to allow for their persistence, greatly increasing the odds for immune dysfunction leading to cancer. Many retroviruses that have affected species in the past have been assimilated into our genetic makeup.

These are known as endogenous retroviruses (ERVs), meaning that they have essentially been disarmed, and we are now living peacefully with them. However, if conditions in the body become abnormal, these viruses can rise up and cause disease.

I recall asking Frank at one point whether the abnormalities I was observing in the cell cultures could allow for a long-dormant bovine leukemia virus (BLV) in cows to start being a problem. Frank was well known in the field as he and Bernie Poiesz isolated the first disease-causing human retrovirus, HTLV-1, a leukemia virus. Frank had published several papers on BLV, and I had done the technical studies. Robert Gallo would essentially get all the credit for HTLV-1, while Frank and Bernie not only got little recognition, but were fired for getting "too much credit." Frank thought my question about immune abnormalities waking up sleeping retroviruses was a valid concern.

I spent a lot of time working on the effect of bovine growth hormone on different cell cultures. It wasn't simply a single experiment.

I took pictures of the cells, showing them swollen and abnormal-looking, how they had become multinucleated, the stringy neurite outgrowths, and how many of them were dying.

I had the readouts of the abnormal molecules, and just as I had reported that the manufacturing process of one Upjohn product would decontaminate HIV, my research on bovine growth hormone was leading me to an opposite conclusion. I always figured a scientist was like an umpire in a baseball game, calling balls and strikes as he sees them.

And that's what led me to be in Russ's office, with the data that the bovine growth hormone would not pass biological quality control. And I could just feel him getting angry, seething inside that I was making his life more difficult, because I was telling him he needed to stop the manufacture of the product, until we could determine whether it was safe. There were other companies, like Monsanto, who were selling similar products, and if we raised the red flag, they would eventually have to follow suit, as well.

I knew it might cause problems, but that wasn't my concern as a lab technician.

The data were the data.

Angry words were exchanged, I stood from my seat, turning to leave, and that's when Russ asked for my notebook and said the words that remain burned in my memory thirty years later, "You have a moral, legal, and ethical responsibility to do exactly as I tell you."

Russ wanted my notebook, and I knew that to some extent he was right. Upjohn paid my salary. The work I had done belonged to them. But I wanted to make a statement.

"You want my notebook? Well, here!"

An old boyfriend of mine, Don Kent, was a Frisbee champion. So instead of simply throwing the notebook at him, I turned it into a Frisbee and sent it sailing just over his head. Russ ducked, moving quicker than I'd ever seen him move. The notebook hit a bulletin board right behind him, knocking off a couple of notes before it fell to the floor.

I stormed out of his office and went to see Wayne, the human resources manager whose office was nearby, and exposed him to the ruckus.

His eyes silently asked the question, "What happened?"

"He's an idiot!" I said quickly.

Wayne laughed. Nobody in the lab liked Russ. "A little more detail, please?"

I told him about the bovine growth hormone experiments, what I thought should be done, how I thought Russ wasn't going to do anything about it, and finally, the Frisbee toss of my notebook at Russ.

"You can't say that, Judy," Wayne said, in response to how I was attempting to tell Upjohn what they had to do. "And you can't throw things at your boss."

"He deserved it."

"Maybe." Wayne was quiet for a minute, and I felt bad for putting him in such a spot.

"It's okay. It's not your problem, Wayne," I said. "I need to go home anyway. Let me call Frank."

The previous day I'd received a call from my mother, letting me know that my stepfather, Ken, had come down with an aggressive type of prostate cancer at the age of fifty-five. She'd wanted to know if there was any way I could come home and help her through what was likely to be a difficult time. The problem was that my stepsiblings had lost their mother to breast cancer, Ken didn't want his children to know about his cancer. If I was going back to Washington, DC, that meant I'd be back in the neighborhood of the National Cancer Institute in Bethesda, Maryland, a distance of fewer than ten miles.

When I reached Frank, I explained what had happened and broached a topic we had put off-limits long ago. Early in my time with Frank, he told me about the abuse he'd suffered at the hands of Robert Gallo, and I told him about a chemistry teacher at the University of Virginia who was adamant that women should never become doctors. He gave terrible grades to any woman in his class, and it destroyed my dream of going to medical school.

After we'd shared our individual traumas, I'd said that we each had our own forbidden "G" words.

I wouldn't say Gallo, and he wouldn't suggest graduate school. But things were different now.

"I'm ready to talk about graduate school," I said to Frank. There were several lab technicians who were working at the National Cancer Institute who were getting their graduate degrees at George Washington University. "Can you get me into George Washington?" I asked.

"I think so," said Frank.

"And can I get my old job back?"

"Probably not. But I can get you some contract work here. We'll figure it out."

Frank was always good to me. It's why we have a consulting business together more than thirty-five years after we first met. Upjohn was good to me, as well. When the next company newsletter came out, they announced with great fanfare that I'd been accepted at George Washington University for graduate school, and they hoped I would consider returning to them in the future.

As for Russ, he was assigned to take sensitivity training for his part in our dispute.

* * *

In September of 2015, researchers from the University of California at Berkeley released the findings of their investigation on the presence of bovine leukemia virus (BLV) in the breast cancer tissue of 239 women. Fifty-nine percent of the breast cancer tissue samples showed evidence of exposure to BLV, while only 29 percent of tissue samples from women who never had breast cancer showed evidence of exposure to the virus.

In a press release from UC Berkeley, the lead author, Gertrude Behring, said:

> Studies done in the 1970s failed to detect evidence of human infection with bovine leukemia virus. The tests we have now are more sensitive, but it was still hard to overturn the established dogma that bovine leukemia virus was not transmissible to humans. As a result, there has been little incentive for the cattle industry to set up procedures to contain the spread of the virus. This odds ratio is higher than any of the frequently publicized risk factors for breast cancer, such as obesity, alcohol consumption and use of post-menopausal hormones.[1]

There's a great deal to consider in that statement. Maybe the tests in the 1970s weren't sensitive enough to detect the presence of the virus in breast cancer tissue. Another possibility is that maybe the cattle during that time weren't plagued by bovine leukemia virus in the numbers that they are today.

Could the use of bovine growth hormone introduced in the 1980s have changed the expression of latent bovine leukemia virus that had been silenced in the genetic makeup of cattle?

Again, I don't know.

But it is a reasonable question and should be pursued with great vigor.

And what do I conclude a person should do when a superior tells them, "You can't say that!" Well, it seems clear to me that in most instances the problem isn't going to go away. I saw clear evidence in the 1980s that bovine growth hormone was affecting cell cultures, making me wonder what effect that might have on the expression of bovine leukemia virus in the milk from those hormone-treated cows.

Today there's a concern that BLV may be linked to breast cancer.

What woman doesn't worry about breast cancer at some point in her life?

I understand that there's a great gap between what I investigated and the research from UC Berkeley, but there shouldn't be. We should understand the entire chain of cause and effect from the introduction of a new product to the public to its effect on human health. I realize the process can be difficult and time-consuming, but we're talking about people's lives.

Science is supposed to be about answering difficult, even unpopular questions. When I first interviewed with Frank, he wanted to hire me but got a thumbs-down from the division manager. When asked why, the woman replied, "She asks too many questions." Frank thought that was exactly what was needed in science and went over her head to hire me.

If you do keep talking when people tell you to be quiet, most times there will be good people like Wayne or Frank to watch out for you, even if you cross a line or two. I've generally found that in whatever situation you find yourself, there will be people who respect honesty and will act to support those who speak the truth.

However, there will still be times when you find yourself alone. There will be no safety net. And you must make the choice of whether to tell the truth, regardless of whether a single person will support you. These situations do not occur often in a lifetime.

They are rare.

But they reveal character.

Like a fireman who decides to run into a burning building to save a child. Or individuals who run toward the sound of gunfire, rather than away from it. Those who decide to help people with deadly diseases. Or someone reporting an abusive boss or company fraud.

These are the decisions that define a person.

If you can speak out about an important issue, when others tell you to be quiet, and you don't know if anybody on the face of the planet will support you, that too, my friend, requires great faith.

Choose wisely.

And if you can avoid it, don't throw notebooks at your boss, even if it's Frisbee style.

CHAPTER THREE

The Dead Doctors — What Is Real?

When I was an undergraduate at the University of Virginia, one of my friends, Jenny, had a t-shirt that read, WHAT IS REAL? It seemed like a cool, sophisticated line to this young college student. But it has taken on a deeper meaning to me over the years. I constantly find myself asking what I really know. I know what people tell me, and what I see on television or read in the newspapers. I'll take from that information a working understanding of the world, but I've learned not to hold those views too tightly.

Take for example my college roommate, Teri. She told me her parents, Thomas and Lucille, were career diplomats. I even rented a room in the house they bought in Charlottesville during our senior year. I felt boring compared to Tom and Lucille, who were worldly, sharing stories of their travels to exotic locations. I loved them and our wide-ranging conversations. I studied my ass off in nutritional biology and chemistry, got up at four in the morning to row crew, spent my days in the lab, and stayed away from marijuana because I knew it was fat soluble, which meant it remained in your body for decades.

Thirty years later I learned Thomas and Lucille weren't diplomats, but agents of the Central Intelligence Agency (CIA). Thomas was CIA station chief in Warsaw, Poland, from 1980 to 1982, just after Teri and I graduated. In November 1981, Thomas and Lucille smuggled Polish colonel Ryszard Kuklinski, a member of the Polish General Staff, from a remote street corner into the United States Embassy, where another CIA team whisked him out of the country. Kuklinski had been an American spy for nine years and carried with him Soviet plans for a possible invasion of Western Europe through Poland. He also carried with him secret documents regarding how the trade union Solidarity would be suppressed if it gained too much power. It was one of the greatest intelligence coups of the Cold War.

If you think I'm sharing national security secrets you'd be wrong, as the heroism of Thomas and Lucille Ryan was prominently featured in the 2004 book *A Secret Life: The Polish Officer, His Covert Mission, and the Price He Paid to Save His Country*[1] and awarded the coveted title of *Washington Post* Best Book. Famed Watergate reporter Bob Woodward called it "the epic spy story of the Cold War."

I also questioned what was real in 1993, when my stepbrother, Kevin, a United States Park Police Officer, was first to find the body of Vince Foster, the Deputy White House Counsel under President Bill Clinton, dead of an apparent self-inflicted gunshot wound in Fort Marcy Park. My stepbrother never saw a gun in Foster's hand, as reported by later eyewitnesses, and a great deal about the scene struck my stepbrother as disturbing.

The bodies of suicide victims who die by gunshot are usually contorted, but Foster was laid out rather casually on the slope of a small hill, as if he'd simply decided to lie down and take a nap. No pieces of his blown-out skull and little blood were found on the small berm, and there were no flash burns from the shot that was supposedly fired from point-blank range at the soft palette of his mouth. There were other curiosities, such as the lack of dirt on his shoes, powder burn particles on other parts of his clothes, a blonde hair from somebody who was not his wife, unidentified carpet fibers, and semen stains in his shorts, suggesting recent sexual activity.[2]

By all accounts, Vince Foster was a straight arrow, but he'd had a long acquaintance with the Clintons. He'd known Bill Clinton since they were in kindergarten together, and he'd worked with Hillary Clinton when they were both partners at the Rose Law Firm in Little Rock, Arkansas. In what was claimed to be a draft resignation letter found torn up in his briefcase after his death, he allegedly wrote, "I was not meant for the job or the spotlight of public life in Washington. Here ruining people is considered sport."[3] The inconsistencies in the story and the evidence spawned many theories, including that Foster had been conducting an affair with the First Lady and become despondent in light of their mounting scandals and killed himself in a secret apartment where they regularly met, or that he'd become a liability to the Clintons and they had him killed.

My stepbrother could offer no evidence for any of these claims.

But he was steadfast in his belief that the crime scene as he first observed it differed greatly from what was in the official report. I saw in my stepbrother's experience how easy it was for a person simply trying to do their job to get swept up in a storm not of their making.

* * *

I was struck nearly speechless in May of 2018 as I stood on the main stage of the Autism One Conference in Chicago, Illinois, in front of more than a thousand people and was presented with the Dr. Jeff Bradstreet Courage in Medicine Award.

It had been a long road, from our groundbreaking 2009 publication in the prestigious journal *Science*, to my firing and jailing in 2011, the forced retraction, the years of my forced bankruptcy, and attempts to obtain a single day in court on my legal claims. After my supposed fall from grace in science, I've been fortunate to meet many other wonderful renegades and rebels, who've given me their love and support. There's a common expression these days: "being red-pilled." It comes from the science-fiction film *The Matrix*, in which the hero is given the option of taking the blue pill and remaining asleep, or the red pill, which will open his eyes to the reality of what is happening in the world.

I'd been red-pilled for maybe eight years, and what a ride it had been. I'd survived the journey, although with a greatly reduced bank account and scientific reputation.

However, Dr. Jeff Bradstreet and many others had not.

* * *

I believe I first met Jeff Bradstreet at a conference in Frankfurt, Germany, in 2012.

Jeff was one of the premier physicians in the United States treating vaccine injury. From the beginning, he understood our approach of looking at autism as an acquired immune deficiency associated with a retroviral infection, similar in many ways to HIV-AIDS. Jeff was accompanied by Drs. Marco Ruggiero, an Italian researcher, and Paul Cheney, whom I'd long known, since my research into ME/CFS. Cheney was one of the first two researchers to identify what came to be known as the Lake Tahoe/Incline Village ME/CFS Outbreak of 1984–1985, which paralleled in many ways the AIDS epidemic.

The three of them were working on a therapy called GcMAF (or Gc protein-derived macrophage activating factor), a protein produced by modification of vitamin D-binding protein, which activates the body's macrophages to fight infection. It was first used as a cancer and HIV therapy. Macrophages are the orchestrators of the immune system response, and what we've learned about retroviruses is that part of their survival strategy is to confuse the immune response. If you can get your macrophages working properly, they will attack the viruses, and healing can begin. That's the way nature intended humans to defend themselves.

Bradstreet, Ruggiero, and Cheney were working with GcMAF and a man from England named David, whom I distrusted from the start. He drank a lot at that meeting and bragged about how much money he was making by selling GcMAF, and that immediately rang my alarm bells. I'm a scientist interested in ending human suffering. I understand people need to make money, but that shouldn't be the thing you brag about.

I was trying to recover my scientific career, having been promised to be included in future studies by Ian Lipkin and John Coffin. They both stated to me there was plenty of evidence for other retroviruses in ME/CFS, but we simply needed to get the validation study out of the way so we could rid ourselves of the VP62 issue. My civil attorney, Dennis Jones, told me that with the bankruptcy I could simply say I'd lost my grants and be free to return to my research on XMRV and other retroviruses associated with ME/CFS and cancer. As you can guess, all those promises came to nothing.

Then in 2013 an invitation came from David to present at a conference in Dubai in the United Arab Emirates. Bradstreet and Ruggiero would also be there. "I'm going to rescue your career," David told me over the phone, offering to pay for my plane tickets and hotel accommodations. That was generous and welcome, as we were confident that GcMAF could be of therapeutic value for the patients, as some of my friends in the United Kingdom had been emailing me with encouraging stories.

Shortly after announcing I was going to attend the conference in Dubai, I got calls and emails from several people with concerns. They said, "Don't go to Dubai. Something will happen." One caller, a friend I'd known for a long time, said, "You're not the only person threatened. They also want to take out Bradstreet and Ruggiero."

I was concerned because I'd also heard from Cheney that in the previous year somebody had cut a hole in his office wall and stolen his computer. Is there really a black market for a doctor's computer? This was sounding suspiciously like Watergate-style corruption.

I didn't cancel immediately.

I called David, trying to learn more about the arrangements, and immediately saw some holes in the security. I'd fly into the airport in Dubai, then be picked up by somebody I didn't know who would drive me for about an hour to the meeting location. I asked David if he could have somebody I knew pick me up at the airport. He did not answer, but a staff member said that was not possible. I called the hotel and asked if they could have a driver pick me up but was also turned down.

I contacted my lawyer, David Follin, and presented him with my dilemma. "You cannot go to lawless places," he warned. "You can't go to Dubai. You can't go to Mexico. You can't go to Nevada. You can't go to India. You will disappear."

That was enough for me.

I called the hotel and cancelled the reservation and emailed David. By coincidence, Lucille Ryan was gravely ill with a blood clot. There was nothing more important than being with Lucille in her time of need.

Yes, I avoided going to a dangerous place in the Middle East by using the excuse that I had to help an ailing CIA agent who had helped orchestrate one of the greatest intelligence thefts from the Soviet Union. These are the absurdities of a life spent around Washington, DC, and in government science. If there is such a thing as the Deep State, I probably know a lot of members.

If they asked for anything, I would be there to help.

This is about all of us.

* * *

In 2014, I was at Autism One again.

I'd first attended in 2010, shortly after the publication in *Science*, which is where I first met my coauthor, Kent Heckenlively.

At Autism One in 2014, there was a special Doctor's Roundtable funded by Claire Dwoskin of the Children's Medical Safety Research Institute and Barry Segal of Focus for Health. We try not to discuss the plague of corruption we all know is swirling around the various chronic diseases and instead focus on the best ways of getting patients better.

The meeting was set to begin in a few minutes when Bradstreet and Ruggiero pulled me aside in the hallway. Ruggiero asked, "In five words or less, why didn't you show up at Dubai?"

"Credible threat on our lives," I said, then paused for a moment before adding, "All of us."

That seemed to send a chill through them.

Bradstreet gave me a big hug and thanked me. They knew I didn't have a dog in this fight. I didn't have an injured child, as Bradstreet did.

Bradstreet asked, "Is there anything we can do to help you?"

"Yeah," I replied. "I've been kind of out of the loop on the recent scientific literature and I can't get to any of the journals and I'm locked out of the medical libraries. Can you help me with getting some access, so I can catch up?"

Bradstreet had just given a talk at Autism One that I'd missed, and he had a flash drive with all of the data on it from recent papers and his own work. He pulled out his flash drive. "Here, this will catch you up."

I downloaded the material to my computer and gave it back to him.

Over the next year I met with Claire, Marco, and Jeff a few times at Jeff's clinic in southern California. I visited with several families, reviewed their protocols with GcMAF, cold light laser therapy, antiretroviral therapies, and cannabis to help repair the damaged neurological system. I believe we also met at least once in Washington, DC, and were often on the phone comparing notes.

Autism One of 2015 was filled with a sense of palpable excitement. Dr. Andrew Wakefield was there with his documentary film crew including Del Bigtree and Polly Tommey. Their film *VAXXED: From Cover-Up to Catastrophe* was based on the devastating revelations of Dr. William Thompson, a senior scientist at the Centers for Disease Control, who revealed that from the years 2001 to 2004 the federal government covered up links between the MMR (measles-mumps-rubella) vaccine and autism, particularly among African American males. Thompson had applied for and been granted whistle-blower status in 2014 but had not testified before Congress. (Thompson still has not testified as of the date of this writing in 2019.)

Bradstreet was very excited, giving the keynote talk, and many of us felt together we had finally broken through to understand the underlying mechanisms, the interplay between the various systems, and how we might effectively intervene for hundreds of thousands of vaccine-injured children. I remember passing Jeff in the hallway as he was on the way to one of his talks and I was on the way to mine and giving him a high-five. We had this. There was no stopping us now.

I expected we would talk again in the next few weeks.

That was the last time I saw my friend Jeff Bradstreet alive.

* * *

On June 17, 2015, the office of Dr. Bradstreet in Buford, Georgia, was raided by federal agents from the Food and Drug Administration (FDA) and the Drug Enforcement Agency (DEA). They were on the scene for several hours.

Kent later conducted a long interview with Jeff's brother, Thomas, about the raid. Thomas said, "I'm sensitive to the word 'raid' because it comes with so much guilt and weight attached to it. It's like a drug bust. They come in, armed with M-16s, they take the money, they take the drugs, and it's off to jail. That's not what happened. They came into his office. They did look at financial records. They took some USB drives. They took some information out of his computers. They didn't lock him down. They didn't take his passport. . . . He wasn't arrested. They didn't seize bank accounts or freeze them. It was just harassment. It wasn't some horrible thing; my life is over."[4] According to Thomas, his brother contacted some attorneys who told him that the worst he was looking at was a fine.

Late in the afternoon of June 19, 2015, a fisherman reported a floating body in a stream that led into Lake Lure, a popular tourist destination in North Carolina. It was Jeff Bradstreet, dead of a single gunshot wound to the chest. If the reports are to be believed, Jeff drove to an inn at Lake Lure to meet his wife. She was to meet him there later after dropping off her autistic son with her ex-husband. On the way up to the lake, Jeff stopped and bought groceries, but when he arrived at the inn, they informed him his room was not ready. He said he'd come back in a few hours and gave the attendant his cell phone number, so he could let him know when the room was available.

We're supposed to believe Jeff then drove five miles to a stream, took out his pistol, waded into the water, then from what appeared to be a nearly impossible angle, shot himself in the chest. In addition, the medical examiner brought in by the family commented that there were no flash burns on his clothing or skin, as would be expected if he had fired the fatal shot. Jeff had been an Air Force pilot and an emergency room physician in a high crime area of St. Louis. He'd fought the FDA a decade earlier in court and won. He was tough.

The only wrinkle in all of this is whether Jeff's family had been threatened. I know that during my baptism of fire, those who were against me came after my husband, David, trying to convince him that I was crazy and dangerous. Turn those closest to you against you. I'm told that Jeff's widow has been very reluctant to help Thomas Bradstreet in the investigation of his brother's death. I suspect she is terrified by what might happen to her children if she speaks out.

When I got the call on my cell phone during a sailing race in southern California, the facts were unclear. First, I heard it was a heart attack, then a drowning, and finally a suicide. After the information came in, I tried to make sense of it, but it didn't add up. I don't believe for a minute that Jeff was suicidal. He was a warrior as well as a strong Christian. If he died by his own hand, it was a sacrifice to protect those he loved.

It seemed like Vince Foster all over again, but the truth is, I didn't have any idea what was real.

* * *

One begins to realize it's dangerous to be a renegade against the establishment.

Strange men following you on your bike and watching your house, and deaths under suspicious circumstances. I'd sleep much better at night if I didn't have to consider these possibilities. But you think to yourself, well, maybe people like me have a renegade streak. One day we're discovering a phantom virus causing a mysterious epidemic and the next we're being raided by the federal government. Maybe there's something to what people like me are saying, and we'll discover what it is in ten or twenty years.

But there's probably a good deal of nonsense, as well.

However, I think it's just as dangerous to be a member of the establishment and have a conscience. Because, you see, I was a scientist with the government for more than twenty years, starting with Fort Detrick, Maryland, and the National Cancer Institute.

I understand the innate goodness of most government scientists, trying to improve the health of humanity. It took me a long time to realize those at the very top probably understand this darkness, the stories they don't want the public to know. But I think there are many working near the top who don't see the darkness.

* * *

During the controversy over XMRV and its relation to diseases such as ME/CFS and autism, we had no fiercer critic that Kuan Teh-Jeang, the editor in chief of *Retrovirology* and second in command to Tony Fauci at the National Institute for Allergy and Infectious Diseases.

At first, we were puzzled by Teh-Jeang's violent reaction to our work.

Frank Ruscetti and I had worked with Teh-Jeang in the early days of HIV research and HTLV-1 (human T-cell leukemia virus), the first human disease-causing retrovirus. Teh-Jeang was an intelligent, charismatic little man always looking for ways to increase the participation of minorities in the biological sciences, which up until that time had mainly been the province of white men.

Teh-Jeang was a top-notch scientist.

However, on December 22, 2010, Teh-Jeang took the unusual step of allowing the simultaneous publication of six negative articles about our XMRV research as editor of the journal *Retrovirology*. I say it was unusual because Frank had talked to several of the reviewers of the papers who told us they had recommended the papers not be published because of significant flaws. None passed peer review. One of the peer reviewers told Frank that in response to these criticisms Teh-Jeang admitted the papers in and of themselves were not up to the proper scientific standard for publication, but "together they make a point and they stand." In a later editorial from August of 2012, Teh-Jeang seemed to take an almost ghoulish delight in the attack he launched on XMRV.

> In this respect, a significant example can be drawn from the six *Retrovirology* papers published in December 2010 that were the first to pivotally correct the then held belief that XMRV was an etiological cause of Chronic Fatigue Syndrome (CFS). In that instance, *Retrovirology's* Open Access format was particularly instrumental in permitting interested individuals who were not career scientists, to freely, rapidly, and fully access those paradigm-changing peer-reviewed publications.[5]

You'd think he was tracking down a murderer with the glee in which he attacked findings barely a year old but that had been vetted by the highest levels of the public health establishment in July of 2009. As I read that section of the editorial, what it said to me is "I can hoodwink other health professionals not familiar with viruses to believe this nonsense because they don't know any better."

You'd be hard-pressed to find anybody to say Teh-Jeang had a habit of acting inappropriately. But that's exactly what he did in 2011 when he stood up in a scientific meeting in Leuven, Belgium, as Frank Ruscetti was getting ready to talk and started shouting, "Stop studying this! Study a real virus! Stop wasting money on this! Wasting money! Wasting money!"

I was in Frank's office at the National Cancer Institute on Monday morning, January 28, 2013, when Kathy came into the room with the news that Kuan Teh-Jeang was dead.

At first it was said he'd died of a heart attack at his office on Sunday night.

Then we heard that he'd shot himself at his desk.

The last story I heard was that he killed himself by jumping off the four-story parking garage at the National Institute for Allergy and Infectious Diseases.

Really?

You kill yourself by jumping thirty feet to the pavement? It might kill you, but there's also going to be a whole lot of pain.

To this day, I have no idea how he died.

* * *

I'd like to share with you a passage from Kuan Teh-Jeang's obituary in *Retrovirology*.

To his friends he was always known as Teh.

> Teh had an infectious enthusiasm and winner's mentality both at work and play. He was a skilled tennis and chess player, a gifted writer, and a great debater with strong opinions on virtually all subjects of science and life in general. Additionally, he had a passion for current events and a love of travel, movies, food, and music.
>
> Teh's death is a blow to the retrovirus research community and we will sorely miss his scientific leadership. He has been central to so much of what we have accomplished together as well as being a supportive and generous friend to many of us individually.
>
> Teh was a friend and colleague to many at BioMed Central, past and present. His passing at such a young age caused shock and great sadness. He was simultaneously one of our most solid friends and supporters, while also being a tireless force driving change and improvement, reminding us not to sit on our laurels, and always ready with forthright but constructive criticism when he thought we needed to do better.[6]

We endorse everything that was said about Teh-Jeang in his obituary. That was consistent with the scientist we'd known for decades. However, it was not consistent with the man who opposed us in the XMRV debate. That person was unhinged.

Here's the only explanation for Teh-Jeang's death that makes sense to me.

I think it's entirely likely that in the middle of the XMRV debate, Teh-Jeang believed what people like Tony Fauci and John Coffin were saying, that XMRV was simply a lab contaminant and had not infected the population in great numbers. These arguments take place on the very edge of scientific knowledge, so when a well-educated person makes a claim and seems to have some reasonable evidence, it's easy to believe them.

Teh-Jeang's obituary talks about how he was a "tireless force driving change" and was always urging people to "do better." Even though he'd savagely attacked XMRV, as the editor of *Retrovirology* he was able to see the latest research, prior to publication. He got to see all the papers. His job was to know things BEFORE other people knew them. Other scientists were discovering different strains of XMRV and how it was associated with disease, which is exactly what one would expect with a newly discovered retrovirus family.

It's likely that XMRV is like Spleen Focus Forming Virus, which needs another helper virus in order to cause damage. Viruses will combine and recombine with other viruses in the vicinity, making it a challenge to identify which virus is causing the problem. Even in HIV, most of the viruses are defective. They're not infectious and transmissible. They're doing their damage to the immune system through other mechanisms.

I believe Teh-Jeang was smart enough to understand this exotic biology, especially when even critics like Coffin were finding that XMRVs might need as little as ten days to recombine with another virus and create a new replication competent retrovirus.[7] These viruses might be bits and pieces of other viruses, which is why if you looked at the virus too narrowly, you'd miss it. Those who really understand viruses know the best comparison of a virus is probably to a piece of computer code. Yes, if you have an entire computer program, like a complete virus, that's probably a more robust system. But you can also have strings of computer code that can mess up your machine. I think in a similar way you often don't need an entire virus, maybe just a few hundred base pairs of a viral envelope, perhaps, to affect immune function. We call these viral particles and act as if they're not harmful, but I think that view is likely to be mistaken.

I think Teh-Jeang was troubled by this. He started to figure it out. At first, he probably kept his doubts to himself. Then he'd start to discuss his concerns with a few close friends, people able to understand the science and the politics, like his boss, Tony Fauci. There are three people I place in what Frank Ruscetti calls the "Unholy Trinity of Science," and they are Harold Varmus, Francis Collins, and Tony Fauci. Whenever you ask yourself why the truth hasn't been told in a critical area of public health, you'll probably find the fingerprints of these men at the crime scene.

How does Fauci respond? Maybe he tries to convince Teh-Jeang that his fears are unwarranted. But Teh-Jeang would persist. He'd want facts. Fauci couldn't provide those, and if he did lie, Teh-Jeang would see through the lies. The system depends on smart people.

But in order to draw those people in, you must also make them believe in what they're doing. It sets up an imbalance if you're going to do something unethical.

Every scientist wants to think their work has integrity. To call a scientist a liar is the worst epithet you can use. I strongly believe an impasse was reached between Fauci and his second in command, Teh-Jeang. What happened after that point? I can't tell you. Was Teh-Jeang shamed by what he'd done? As a Chinese man, honor was very important to him.

Did he do things of which he was ashamed?

Or did Fauci, after it was clear to him that Teh-Jeang could not be turned, place a call and give an order? Was Teh-Jeang getting ready to go rogue? Fauci might not have known what exactly would happen, but he knew something would happen.

The rumor is that Teh-Jeang left a suicide note, which was confiscated by the National Institutes of Health police.

I wonder if it resembles the torn-up note found in Vince Foster's briefcase.

* * *

How long have suspicious deaths taken place in government science? I wish I could say it was a new phenomenon. I suspect it is not.

As I said earlier, from 1983 to 1986, I worked with Frank Ruscetti in the Biological Response Modifiers program at Fort Detrick, Maryland. It was the early days of HIV research, and I was a technician, spending most of my time in a Bio-Safety Level 3 facility, which is probably a good thing when all the chemical and biological weapons that had been used and were stored on the base were later revealed. The building I worked in had once been used to test weaponized anthrax. It was a charming place.

I caught a senior scientist changing his data and reported it to the program director. I'm the one who got chewed out. That incident is what made me stop working for the federal government and spend a brief amount of time working in the corporate world at Upjohn.

As I look back on my time at Fort Detrick, several things become clear to me.

In 1985, we were in the heat of HIV-AIDS research, with everybody knowing that Nobel Prizes were on the line, billions of dollars from treatment protocols, and millions who were angrily calling for us to find an answer to their terminal disease. I remember going to meetings several times and having to pass though ACT-UP protests where people were shouting at us that we didn't care about them and were lying about what was really making them sick. I was in my midtwenties and in my naïveté, believed we were doing everything possible.

Sodium azide is a nasty chemical, which uncouples your respiratory chain, so you can't make oxygen. You literally drown in your own secretions. One female technician took some sodium azide from the lab, went to a local park, sat on a rock, and drank it, killing herself. And I've already mentioned the postdoctoral student, a Japanese man with a wife and two kids, who I was told had been instructed to falsify data. He also died from drinking sodium azide.

They eventually banned sodium azide from the lab.

I have nothing but respect for the intentions of people who want to pursue a life in science. My anger is with those who betray the search for truth. I think it's very plausible that honest people stumbling upon corruption can become so despondent that they kill themselves.

But I also think it's likely that procedures are in place to deal with those who are not willing to go along with the program. I'm not saying there haven't been times in my ordeal where I haven't been very despondent. However, for the record, I will never kill myself.

* * *

Do these suspicious deaths of leading doctors and scientists continue to this day?

They do.

In January of 2018, a writer named Baxter Dmitry published an article in which he claimed a senior scientist at the Centers for Disease Control told him, "Some of the patients I've administered the flu shot to this year have died. I don't care who you are, this scares the crap out of me. We have seen people dying across the country of the flu, and the one thing nearly all of them have in common is they got the flu shot."[8]

On February 12, 2018, Dr. Timothy Cunningham left his job at the Centers for Disease Control early, claiming he didn't feel well. If we are to believe the news accounts, after arriving home, he put on his black jogging shoes, went for a run, and disappeared.

Dmitry then revealed that his anonymous CDC doctor was none other than Dr. Timothy Cunningham. When NBC News reported on the mysterious disappearance a week later, this is what they had from Joe Carlos, a friend of Cunningham's from Morehouse College. The two were going to attend a gala at Morehouse:

> "Our last communication the week prior was about hanging out before and going down to the VIP reception and enjoying ourselves," Carlos said. "I can speak for myself and so many classmates that this is very, very shocking."
>
> Cunningham's friends described him as opinionated, positive, and happy—and they noted his reliability.
>
> "He has this pristine service record and background, and then he's also the guy you can call to help you move furniture or get together with you at a restaurant at the end of a long day," said Calloway, who also knew Cunningham from college and maintained a close relationship with him over the years.[9]

In October of 2017, Dr. Cunningham had been named a "40 under 40 Award Honoree" by the *Atlanta Business Chronicle* and sat for an interview with them.

When asked about his work he said, "I'm very fortunate to love what I do. That does not mean it is always easy, but my passion for the work helps sustain me when it gets hard. My advice is to do what you love. Love what you do. Do not quit. Keep going. During the bad, pick yourself up and learn from it. Finally, take time to celebrate during the good times."[10]

Cunningham was also asked what skills and attributes a person under forty can bring to a workplace to make it grow. He replied, "Be flexible. Move beyond your traditional silos. Be aware of your inherent biases. Be open-minded and willing to learn. Open yourself up to get to know people different from you and let them know you as well." Sounds a lot like my approach to science.

On April 3, 2018, more than seven weeks after he'd disappeared, some fishermen in the Chattahoochee River spotted a body tangled in debris. It was Dr. Timothy Cunningham.

In his final conversation with his sister, Tiana, Cunningham had told her "You have to figure things out for yourself."[11] Tiana later recalled that her brother sounded paranoid.

In an extensive article for the *Atlanta Journal-Constitution* on June 4, 2018, a new narrative about Timothy Cunningham began to emerge:

To those who knew him best, Timothy Cunningham was a well-educated, motivated career man who felt it was his purpose to change lives. But in his personal life, Cunningham struggled with his sexuality, was upset he hadn't landed a promotion, and lived with a chronic disease, family and friends told Atlanta police.

It's now no longer a mystery how the Centers for Disease Control and Prevention epidemiologist died. Cunningham committed suicide by drowning himself, according to the Fulton County Medical Examiner's office.[12]

Maybe.

I just don't buy it, though. Especially after what I lived through in 2011. The events were so similar it made my skin crawl. I shudder to think what might have happened if we did not own baby Jonah, that thirteen-foot Boston Whaler. And what about Bradstreet, found dead in a river, as well? Cunningham was well liked, accomplished, had two degrees from Harvard, and was a commander in the US Public Health Service, having responded to the Ebola and Zika outbreaks, as well as Superstorm Sandy. I challenge you to look at a picture of this well-built thirty-five-year-old African American man in his commander's uniform, standing in front of an American flag and the flag of the US Public Health Service, and see a brittle soul.

Do you see how the *Atlanta Journal-Constitution* article muddies the waters?

It's like one of those horoscopes where they say so many things you think it's accurate because one of them resonates with you.

He didn't know if he was gay or not! That's why he killed himself?

He didn't get that promotion! That's why he killed himself?

He had a mysterious unnamed "chronic disease" for which he was taking medication! That's why he killed himself? Okay, what was that chronic disease? Psoriasis?

I tend to think the answer resides in something vague reported in a *Washington Post* article a few weeks after Cunningham disappeared:

Police investigators are bewildered as they work through the "extremely unusual" circumstances surrounding the missing-person case of Timothy Cunningham, a researcher who vanished February 12, shortly after hearing why he had been passed over for a promotion at the Centers for Disease Control and Prevention (CDC).

Cunningham, 35, told colleagues he was not feeling well and left work at CDC headquarters in Atlanta, not long after speaking with his supervisor about why had not been promoted, Atlanta Police Maj. Michael O'Connor told reporters.[13]

I'd like to know the reason Cunningham was passed over for that promotion.

Was it because he was saying flu shots were killing people?

That might get you passed over for a promotion in public health if it turns out you're killing members of the public and talking about it to nosey reporters.

* * *

How far will our government go to attack members of what it perceives to be its domestic opposition? In other words, what measures will the United States government employ against its own citizens? This isn't a question of one party against another. It's a question of power.

I tend to think that looking at the past can help us answer such questions.

If somebody stepped over a line, we expect that the person will be punished.

If it was part of a standard operating procedure, we can expect that the identity of such persons will remain hidden, even if the crimes are exposed.

On November 21, 1964, the Federal Bureau of Investigation (FBI) sent a letter to civil rights leader Martin Luther King, Jr., urging him to kill himself, along with a tape recording of his alleged sexual encounters with women. King was supposed to be traveling in a few weeks to Sweden to accept the Nobel Peace Prize for his work aimed at ending segregation.

In 2014, the *New York Times* printed a nearly unedited copy of that letter. The leadership of the FBI apparently wanted King to believe it came from a fellow African American.

KING,

In view of your low grade, abnormal personal behavior I will not dignify your name with either a Mr. or a Reverend or a Dr. And, your last name calls to mind only the type of King such as Henry VIII and his countless acts of adultery and immoral conduct lower than that of a beast.

King, look into your heart. You know you are a complete fraud and great liability to all of us Negroes. White people in this country have enough friends of their own but I am sure they don't have one at this time that is any where your equal. You are no clergyman and you know it. I repeat you are a

colossal fraud and an evil, vicious one at that. You could not believe in God
and act as you do. Clearly, you don't believe in any personal moral principles.

The next three paragraphs spew hate and attack Dr. King's character and
Christian values, mocking his degrees and honors. Once again, this type
of dialogue is eerily similar to the kinds of things written about me. The
removal of the Dr. from my name and my entire life's work seemingly stolen
as others like Coffin and Lipkin appeared to cash in on our discoveries by
publishing our data and patents as their own. The FBI and other federal
officials ostensibly participated in all of this in an attempt to convince me
to commit suicide in exchange for destroying my family and my honorable
name. Am I surprised how the letter ends? Decide for yourself.

> King, there is only one thing left for you to do. You know what it is. You have
> just 34 days in which to do (this exact number has been selected for a specific
> reason, it has definite practical significant.) You are done. There is but one way
> out for you. You better take it before your filthy, abnormal fraudulent self is
> barred to the nation.[14]

Does that letter sound as psychotic to you as it does to me? This is our
Federal Bureau of Investigation urging the country's most prominent civil
rights leader to kill himself prior to accepting the Nobel Peace Prize.

In fairness, it must be noted that Dr. King did not lead an exemplary
personal life.

Neither did President John F. Kennedy, or President Lyndon Johnson,
who followed him.

But the FBI never sent letters to Kennedy or Johnson urging them to
commit suicide.

I know many will see that letter from the FBI to Dr. King solely through
the prism of race relations in the 1960s. I think it's about something even
larger. If the use of this tactic was an aberration, then I'd expect that the
person who wrote this letter would have been identified, stripped of his job
and pension, and criminally charged.

Nothing like that even remotely happened.

The author of this hideous letter was allowed to recede into obscurity
without penalty, just as to this day no one has ever paid for the crimes committed against me or this patient population. In fact, they are awarded with
tens of millions of dollars in federal grants.

It's a larger question than racism or the corruption of science. It's the question of who's allowed to have a voice. Who is permitted to participate in the conversation?

Dr. Martin Luther King, Jr. was assassinated on April 4, 1968, in Memphis, Tennessee. I remember the day, as it was only days after my tenth birthday. I remember wondering what it was like for King in those final months, as my grandfather said he knew that bullet was coming for him. He'd seen the shadow of its trajectory during his marches and rallies, as he sat in Southern jails, and when that letter arrived from the FBI along with the tape recording.

Soon they would come for him.

It was only a matter of time.

And yet he persisted.

We must do the same.

CHAPTER FOUR

The Fate of Those
Who Fight the Darkness

As my attorney, Mike Hugo, tells it, he first heard about me when he got a call from an attorney in Southern California with whom I'd discussed the case.

The attorney said, "Hey, Mike, I've got this case. It's probably the hardest you've ever seen. And I don't know if the case is hard because the client is so emotional and tortured by the factual circumstances that she has trouble articulating herself or what. She's really brilliant. But what happened to her is so off the wall that the first person I thought of was you. Because you like off-the-wall stuff."

"Okay," said Mike, finding himself intrigued by that opening. The other attorney laid out what he knew of the case, my background, how I was arrested and jailed for five days without an arrest warrant, then let go as if by magic when my former boss, a wealthy political donor to US Senator Harry Reid, told the police to let me go.

Mike got all the facts he could from this lawyer and then gave me a call. We talked for hours. I can be like that. Sorry, Mike.

Our book *PLAGUE* was in galleys, so I sent him a PDF copy. I also sent him a copy of the 42 USC 1983 complaint I'd filed *Pro Se* (on my own behalf, from the Latin) with the court. I hadn't been able to find an attorney prior to filing the complaint, but a few of my friends who were attorneys gave me some general guidance. The problem was they weren't trial attorneys, and the facts of my case were rather unusual.

Mike found the complaint to be poorly drafted and confusing. I agreed. I was a scientist, not a lawyer. We were facing a motion to dismiss my complaint, and Mike did a wonderful job in front of that judge.

As Mike later explained, "A poorly drafted complaint usually has a lot of facts in it. If you write a good complaint, it should come down to three pages."

When Mike appeared in front of the judge, he simply laid it all out on the line. "I'm totally confused. I'm asking the court not to ask me any questions today and let me redraft the complaint. I need to talk to my client and figure out what this case is really all about."

The judge seemed relieved by Mike's candor and granted him the time he needed to amend my complaint.

* * *

There's probably no better attorney in the entire world to understand what happened in my case than Mike Hugo.

Mike got his undergraduate degree from Boston College in 1975, then graduated magna cum laude from New England Law School in 1983. His first few cases out of law school involved vaccines, specifically the old DPT (diphtheria-pertussis-tetanus) shot, which was eventually removed from the market because of so many bad reactions. Within a year of doing that first case, he had more than a hundred vaccine cases and at one point was handling somewhere around five hundred DPT injury cases. At the time, he was also developing a practice based around women who were alleging health problems from their silicone breast implants.

An attorney friend of Mike's was doing an environmental case and suing two large companies, W. R. Grace and Beatrice Foods. It would come to be known as the Woburn case, eventually resulting in one of the largest environmental fines ever levied against a corporation. The companies were trying to spend that attorney into the ground, and Mike realized that he could use the positive cash flow he was getting from his vaccine injury and breast implant cases to help that attorney.

That attorney was Jan Schlichtmann, portrayed in the 1998 movie *A Civil Action* by John Travolta, and in the popular book of the same name. Jan has famously said, "John Travolta made more money playing me than I ever did playing me."[1]

That's something of an understatement.

At one time Schlichtmann had $114 in his bank account and a million dollars of debt. Mental breakdowns are common among those who try to change the world. Jan was no different, eventually fleeing Boston after the loss of the Woburn case to Hawaii and living in the crater of Haleakala, where he began selling energy-efficient light bulbs.[2] It was only after Jan's loss in the Woburn case that the federal government would conduct a thorough investigation of the claims, leading to the record fines.

During the years Jan was preparing the Woburn case, he was able to do it because Mike was bringing in money with his vaccine injury and breast implant cases. The Woburn case left Mike with many financial obligations, some of which continue to this day.

* * *

The vaccine injury cases were lucrative to Mike's practice, but exhausting.

These were in the days before the passage of the 1986 National Childhood Vaccine Injury Act that would change everything. Mike said there were about eight attorneys around the country who were handling these cases and they would regularly get together to meet and discuss strategies and tactics. Their main experts were Dr. Kevin Geraghty, a physician from Northern California who had completed fellowships in allergy and immunology at the University of Chicago and the University of California, San Francisco; Dr. Arthur Zahalsky, who taught at City University of New York and was chair of the Department of Biological Sciences at Southern Illinois University-Edwardsville; and Dr. Mark Geier, who had been an obstetric geneticist at the National Institutes of Health.

As Mike recalled, "We'd sit and compare notes about our cases, what their experts were saying, and prepare one another for what we were going to do. That evolved into analyzing one another's cases and offering help. Expert witnesses we had, that the other guy might not. Because we were all in it together. We all needed to win. We couldn't afford a loss because we needed to show how bad this vaccine was."

The vaccine litigation was vicious and exposed Mike to a level of bad behavior that he hadn't believed was possible. The worst was a cluster of eleven sudden infant deaths in Tennessee due to the DPT vaccine manufactured by Wyeth Laboratories. The truth was uncovered by a state medical examiner who asked the right questions in the medical histories of the eleven infants who died.

In discovery, Mike uncovered an August 27, 1979, memo from Wyeth, which was sent by the company's head of vaccines, a Dr. Alan Bernstein. While the number of deaths caused by this vaccine was shocking, the number of brain-damaged children was even higher. Those were the cases Mike was handling in the early 1980s.

The memo started out with the acknowledgment the vaccine might kill a certain number of children, which was why they decided that no more than two thousand doses of the drug would go to any metropolitan area, thus avoiding the situation where many children would die in a small geographic area in a short period of time.[3] This way you might get one or two infants dying in Boston, one or two in New York, a few in Philadelphia, a few in Nashville, but not enough in any single area to cause a panic. Babies would die, or be grievously injured, but to their parents and the local health authorities, what happened would remain a mystery.

At least, that's the way Mike Hugo interpreted it.

When Mike deposed Dr. Bernstein, he went after him with a vengeance. "I ran this guy over the rocks as hard as I possibly could. I blamed deaths on him personally. I said there was blood on his hands, not on Wyeth's. I could not have been a bigger asshole. The next morning his lawyer tells me, 'Dr. Bernstein was in the hospital all night with chest pains. I hope you're satisfied.' I replied, 'I hope Dr. Bernstein is satisfied that watching your child go from normal to brain-damaged or dead is more painful than a heart attack.'"

The pace was exhausting. Mike would often fly out of Boston on a Sunday night to a variety of different cities—New York, Dallas, Chicago, Los Angeles—then fly back to Boston on a Friday night, rest up for a day and a half, then get ready to do it all over again. One night he found himself awake on the floor of a Marriott Hotel room in Salt Lake City where he was staying for a week for a deposition, crying because he didn't know where he was or why. He had to crawl over to look at the address listed on the room phone to realize where he was.

"What the hell am I doing in Salt Lake City?" he thought, before he realized he'd already been there for two days of depositions and had three more to go.

Mike had all these cases going on across the country when he got a call from a lawyer in Washington, DC, whose child had been brain-damaged by the DPT vaccine. He said he'd been through the litigation process and they'd won, but it had been too difficult.

Something else needed to be done.

Vaccine cases were taking about six years to judgment at that time, costing about a hundred and fifty thousand to prepare. Mike recalled, "You get to the point that you know enough to make the decision yourself. You don't need to spend twenty-five thousand dollars on a pediatric neurologist and a geneticist. If the child was born normally, everything was on target, and all the milestones were made and suddenly you've had a loss of function within twenty-four hours of a DPT shot, I know enough to file a lawsuit. But once you file the lawsuit, you've got to take all the depositions. You have to analyze all the records. You have to know there's something peculiar about one of the records. You have to stay in hotels where sometimes you wake up in cold sweats on the floor. And it costs money to do that."

But often the injury wasn't immediately apparent within twenty-four hours. Yes, there could be death, there could be paralysis where your kid was suddenly a spastic quadriplegic, but other injuries were subtler. "In some cases, the kids are going to school and the teachers are having a conference with the parent and say, 'Your kid is not doing what the other kids are doing.' And in 99 percent of the cases the parents break down in tears because they knew it all along but didn't want to face it. But now a teacher is telling them and now they have to do something. And that's when they call a lawyer like me."

The lawyer in Washington, DC, was talking to some people in Congress about how to create a more efficient and less painful system to get compensation to the parents of vaccine-injured children and wanted to get Mike's support. Mike thought that any change to the system would be an improvement.

Mike spent hundreds of hours working with the office of Congressman Henry Waxman and his legislative aide, Tim Westmoreland, trying to craft a worthwhile piece of legislation. He is unsatisfied with what has happened with that law.

"There has to be an out," says Mike, speaking of the 1986 National Childhood Vaccine Injury Act. "If parents are willing to go through the process of proving the product was defective, you have to allow them to do that. They have to be able to get a ruling in 240 days or they can opt out. The law was supposed to get those parents a ruling in 240 days, so they could get back to taking care of their injured child. Now it takes three or four years and we're back to where we started. It's a crazy system. It has to change."

* * *

Mike's years as a plaintiff's attorney have also opened his eyes to another unsettling development of which I venture to say most of the public is unaware: corporate surveillance.

In one of his cases he was representing clients from Arkansas in an opi-oid case. His clients lived on a five thousand-acre farm, and their home was at the intersection of two roads. On one corner was the client's house, mom and dad lived in a house across the street, on another corner was the barn, and on the other corner was a big garage where they kept their tractors.

Imagine this rural Arkansas town, then picture a black Ford Crown Vic stationed at the intersection, just sitting there for days on end. When the client phoned Mike and told him about the car, Mike immediately knew who it is was.

Mike called counsel for the other side to poke fun at the surveillance. "At least have your guy rent a tractor and go to Sears to buy a bunch of wrenches. He could pretend to be fixing his tractor. He'd blend in a lot better than a guy in a Crown Vic."

"We can legally follow anybody we want," the lawyer brusquely replied. "We could even follow you."

"Yeah, if you could catch up to me," Mike joked back.

A few days later, Mike needed to drive from Boston to Portland, Maine, for a deposition. Another one of the lawyers was handling a case where a child had drowned in a swimming pool at a resort in Maine. He asked if Mike wouldn't mind going to the resort and taking a few pictures of the pool and access points. Mike said he would go.

Before leaving his house that day, Mike had filled up his BMW X-5, a sports utility vehicle with a thirty-gallon tank that gave him about a five-hundred-mile driving range. He'd been driving along Interstate 95, a major road, when he noticed what he thought was a Chevy Escalade. It had an unusual arch from the cab to the bed, and he thought it looked sharp. He paid the toll on the highway to get into New Hampshire and noticed the truck is behind him, and the same thing when he paid the toll to go into Maine.

Mike decided to speed up to see if the Escalade would follow, even as he approached a hundred miles an hour on a desolate stretch of road.

The Escalade followed.

Then Mike decided to slow way down to about thirty miles an hour.

Still, the Escalade stayed behind him.

"Okay, I know what's going on," Mike said to himself. "Let's have some fun."

Mike sped past Portland on his way to the resort where he needed to take pictures, and the Escalade followed. Mike took his pictures, then got back into his car. He'd grown up in this area of Maine and had a couple hours before his deposition. He also knew a good fifty-mile stretch of road without any gas stations and figured the henchman must be running low on gasoline.

Mike headed northwest on a desolate stretch of road, and sure enough the Escalade ran out of gas. Mike watched in delight as the driver pulled onto the shoulder of the road. Eventually Mike turned around, so he could pass the stranded driver. As he drove slowly past, he flashed the guy a big shit-eating grin, honked his horn, and gave the guy the middle finger.

This was when cell phones were relatively uncommon. Mike had one, but he didn't know if the surveillance guy did. Mike called the attorney in California and said, "Your guy couldn't keep up with me. He ran out of gas." Mike gave the attorney the approximate location and said, "You might want to send a tow truck."

There were other incidents, less terrifying maybe than having a car on your bumper as you're barreling down a highway at a hundred miles an hour, but no less disturbing.

Traveling across the country and taking depositions is a lonely life, and as attorneys are for all practical purposes warriors in the cause of their clients, there's often a mutual respect that develops between the combatants. You may fight during the day in a deposition with all your brains and skill, but at night you're just a couple guys stuck in a hotel far from family and friends.

In a case against a manufacturer of Rezulin, a diabetes drug, which was causing kidney and liver failure, Mike was getting ready to do a deposition, and the lawyer for the other side asked where he was staying. Mike replied he was a Marriott guy, so he'd be staying at one of their local hotels.

The attorney replied, "We're all staying at the Chesterbrook Embassy Suites. Why don't you stay there? When the day's over we can have a drink, maybe dinner, and go over the day?"

Mike thought, "What the hell? Who could be hurt by that?"

One night, as he was in the middle of a week of depositions, he was talking with one of his partners about a series of studies that he planned to bring up in the next day's examination. However, during the next day's deposition, the witness seemed to be too familiar with the studies he was bringing up.

Mike called for a break, then went to his hotel room. He went to the phone, unscrewed the mouthpiece, and saw that there was a listening device in the receiver. He took pictures, replaced the mouthpiece, and checked out of the hotel.

A few weeks later, Mike got a call from one of his fellow vaccine lawyers who also had a case against Wyeth. The attorney was asking Mike if he had a certain document that might help in his case.

"Where are you staying?" Mike asked.

"Embassy Suites," the attorney replied.

"Let me take the mystery out of it," said Mike. "You're staying in Room 585."

"Yeah, how did you know?"

"Hang up the phone and call me from an outside pay phone and I'll tell you."

Now, some people may think this is something that remains in the bad old days of the past, but Mike claims he continues to hear stories like this from attorneys working today. One of his friends, who was pursuing a case against General Motors, claims the General Motors attorneys were listening to his conversations through the On-Star system.

Kent is deeply disturbed by these allegations, seeing this surveillance as akin to the mafia listening in on the federal government with impunity. The public is being harmed because these corporations are acting in an extrajudicial fashion never envisioned by the framers of our judicial system.

Mike sees the situation differently: "I think their goal is to protect their corporation from that first Vioxx verdict, which was $242 million dollars. Or that first breast implant verdict, which was forty something million dollars. Because they must do everything they can to protect their investments. I'm not sure the goal of the surveillance is to undermine public health and safety. It's to undermine the lawsuit you want to bring and taking their money. We're talking serious money. You can pick up the newspaper and read about the recent six billion-dollar settlement with Johnson and Johnson for asbestos in baby powder talc. That's one you can watch right now. They must defend their money. If you ask them, there's nothing wrong with what they're doing. They're protecting their investors, their investment, their board, and that's what it's about."

I'll let you come to your own conclusion, but it seems like these are the two sides in a critical debate about which we all need to have an opinion.

* * *

One of the most surprising things about Mike Hugo is how he can see the flaws of his own side, as well as those of the corporations with which he often finds himself in dispute.

Probably the greatest challenge a products liability lawyer has in bringing a new case is something called the Daubert standard. In theory, it all makes sense. The legal system wants to ensure good science is presented in the courtroom. Lawyers are not scientists. However, there is much more money provided by corporations to scientists wanting to show the safety of their products rather than their danger. I don't think there's a single person who can reasonably dispute that claim.

If you are making a new claim regarding the danger of a product, you need to be sure you have some solid science. Because once a decision is made under the Daubert standard, it may be several years until a court will let you reopen the issue. There you are, as the attorney with these clients who believe a certain product harmed them, and you also believe it to be true based on the evidence you've reviewed but worry about being able to prove it in court. There are more scientists with their hands out to do research for corporations than scientists who will dig into their own pockets to fund safety research.

You need to wait until the science is extremely strong, and if you don't, you're going to screw things up for other people for years to come.

At one point, Mike represented eighteen hundred women who had health problems related to their silicone gel breast implants. The science was strong on the development of health problems if their breast implants leaked. But Mike was also finding serious health problems in women whose breast implants had not leaked. There were autoimmune issues like Reynaud's Syndrome, lupus, and serious skin conditions where the tissue was attacked and hardened, and you might lose all the skin on your body.

"I had one videotape sent to me by a woman, and they actually had her on a bed with all these misters like you have in a garden because she literally had no skin on her body," Mike recalled. "She didn't have any eyelids. She had no skin on her body. And they grafted a hundred percent of her body and she survived. It was like a miracle. And that was from silicone."

However, there was a lawyer from Portland who wanted to bring a case that the silicone gel breast implants were causing disease, even though they weren't leaking. Unfortunately, the science wasn't ready. It was close, but not there yet. Several researchers were doing great preliminary work and were probably about six to eight months from being able to publish in a top journal like *The Lancet*. But this lawyer wouldn't be dissuaded.

He brought his action and lost, screwing up all those breast implant cases.

For many years after that, if anybody wanted to win a silicone gel breast implant case, they had to allege that there must have been a leak of the material, even if there was no evidence to support it.

The same lawyer came to Boston a few years later to talk with Mike. He said, "I've got another theory on a different piece of litigation, and I need your buy-in on it."

"What is it?" Mike asked.

"Vaccines cause autism."

"You're not going to try to prove it in those terms, are you?"

The lawyer said, "Yeah, I think it causes autism and we're going to prove it. We've got this guy from England who's going to testify, and we've got a ton of information."

Mike knew the lawyer was talking about Dr. Andrew Wakefield and his research on the MMR (measles-mumps-rubella) vaccine and the development of gastrointestinal problems, as well as autism.

"Yeah, you're right. Vaccines do cause autism," Mike replied. "But you can't go into court and prove that for two reasons. First, the government will never allow that to happen. And number two, the science isn't there yet. I know about Wakefield. All his work describes to me is a postpertussis vaccine encephalopathy, which is compensable under the vaccine program. So, talk about it as an encephalopathy, which is the same fucking thing. But just don't use that word, 'autism.' Never say the A word. Never do it. Never do it."

"No, you don't understand," the lawyer protested.

Mike's emotions were at a fever pitch. "No, *you* don't understand! You fucked up breast implant litigation, and now you're going to fuck this up!"

But of course, the lawyer didn't listen to Mike and the vaccine-autism cases were lost in the Vaccine Court.

I understand Mike's point of view and even have some degree of sympathy for it. He's trying to get the best result for his clients in a corrupt system.

However, when a person stands and swears an oath in court, he promises to tell "the truth, the whole truth, and nothing but the truth." There's no provision for "I'll tell as much of the truth as I can get you to believe."

* * *

And speaking of failure, let's talk about my inability to get even a single day in court for the violation of my civil rights for my false arrest, imprisonment, and how after more than seven years I still don't have any copies of my notebooks, or those of my research team, for the nearly five years I worked on ME/CFS and retroviruses, or in fact for my entire research career.

All of my work was in my two offices at the University of Nevada, Reno campus that was locked down to me and my staff on September 29, 2011. I don't even have a copy of my doctoral thesis on HIV!

I'm an American scientist, for God's sake! Give me my constitutional rights and my data!

Mike has a wonderful perspective on this issue, and his story leads back to Fort Detrick, Maryland, and one of my early jobs in science, working as a young protein chemist purifying interferon in the Fermentation Chemistry lab and doing immune therapy for cancer and AIDS at the Biological Response Modifiers program with Frank Ruscetti in the early 1980s.

The government doesn't like to talk about the fact that in the past they've had a very robust development program for biological and chemical weapons. However, in November of 1969, President Nixon gave a speech in which he announced an end to the American program of offensive biological weapons and reaffirmed a no-first-use policy on chemical weapons.[4] The short document, barely more than a page, is very illuminating. On biological weapons, the section reads:

> Biological weapons have massive, unpredictable, and potentially uncontrolla-
> ble consequences. They may produce global epidemics and impair the health
> of future generations. I have therefore decided:
>
> -The United States shall renounce the use of lethal biological agents and
> weapons and all other methods of biological warfare.
> -The United States will confine its biological research to defensive mea-
> sures such as immunization and safety measures.
> -The Department of Defense has been asked to make recommendations as
> to the existing stocks of bacteriological weapons.[5]

On the issue of chemical weapons, the mandate was a little less clear. Were we stopping our research, or just putting a little fancy window dressing on the controversy?

You decide:

> As to our chemical warfare program, the United States:
> -Reaffirms its oft-repeated renunciation of the first use of lethal chemical
> weapons.

-Extends this renunciation to the first use of incapacitating chemicals. Consonant with these decisions, the Administration will submit to the Senate, for its advice and consent to ratification, the Geneva Protocol of 1925 which prevents the first use in war of "asphyxiating, poisonous or other gases and of bacteriological methods of warfare." The United States has long supported the principles and objectives of this Protocol. We take this step toward formal ratification to reinforce our continuing advocacy of international constraints on the use of these weapons.[6]

Did you catch what was really in that statement?

The Geneva Protocol of 1925 banned the first use of chemical and biological agents in warfare. Nixon was finally submitting this protocol to the Senate in 1969. That's a span of forty-four years, taking in the Depression, World War II, the Korean War, the Kennedy assassination, and a good chunk of the Vietnam War. During the Korean War there were claims by the North Koreans that we used biological warfare against their troops. Did Nixon's statement deny that the United States had ever used such tactics? And his statement still left open the use of deadly herbicides like Agent Orange, of which we dropped massive amounts in Vietnam.

Was Nixon declaring the United States was banning research into chemical and biological weapons so that the world could rest easy that such threats would never come from the United States? Or was there a loophole, which would allow these programs to proceed?

Neither our association with the Convention nor the limiting of our program to research will leave us vulnerable to surprise by an enemy who does not observe these rational restraints. Our intelligence community will continue to watch carefully the nature and the extent of the biological programs of others.

These important decisions, which have been announced today, have been taken as an initiative toward peace. Mankind already carries in its own hands too many of the seeds of its own destruction. By the example we set today, we hope to contribute to an atmosphere of peace and understanding between nations and among men.[7]

As I read the statement, it seems to be business as usual. This public acknowledgment, however, was important, and I genuinely believed it was a first step to a better future.

One needs to tell the truth before a solution can be found.

Fort Detrick was changed from a chemical and biological weapons research lab to a cancer research lab. A decade later, Frank Ruscetti and I came on the scene. We were never hired to figure out ways to end life. We were hired to learn how to save lives. Yes, I know the same knowledge can be used in both pursuits.

But Frank and I have always been about protecting life.

However, Fort Detrick couldn't quite escape its past as a chemical and biological weapons lab. I mean, where were they supposed to store all these biological and chemical agents? Fort Detrick became the main storage facility for these programs. And in fact, the army used the area around Fort Detrick to test aerial concentrations of Agent Orange, using helicopters and airplanes.

As Mike later explained, Fort Detrick had all the terrible chemicals he'd encountered in the Woburn case, as well as biological agents. It was a chemical and biological waste dump.

* * *

Mike came to learn about Fort Detrick because of a case he brought for a man named Randy White, a famous televangelist preacher whose congregation was called "The Church Without Walls."[8] Randy lived next to Fort Detrick. His wife and daughter both came down with cancer and died, which Randy blamed on toxins from the base. His website notes his founding of the Kristen Renec Foundation in memory of his daughter, as well as the Fighting for Frederick project, which is "initiating effective governmental legislation surrounding chemical contaminants and the effect on our population and wildlife." During the Bill Clinton presidency, Randy served as an "official counsel" to the president and was granted top-secret clearance with the US government and State Department.

When Mike looked at a map documenting cancer deaths around Fort Detrick, it looked extremely disturbing: "This map had a black dot for every house in which there was a cancer death. The map was almost black. There were houses where mom, dad, and all three kids died of cancer."

It was difficult bringing the suit in an area with a very promilitary mindset.

But to Mike it wasn't a question of being pro- or antimilitary.

In fact, most of his clients were ex-military or military families because of the common practice in the South that veterans often like to settle near where they served and had some of the best times of their lives.

There was another similar case working through the courts, in which a real estate developer found that he couldn't build houses on land he bought because of the contamination.

The judge hearing the case was a Republican appointee, and Mike thought it might result in a favorable outcome, considering that in a fight between a businessman and the government, the conservative judge would rule in favor of the businessman. However, the judge ruled in favor of the government.

The only additional piece of information Mike had that he thought might be helpful was Nixon's 1969 executive order. As Mike later recalled, "The first thing this judge asked me about the executive order was 'How much of this information you're showing me is public information?' I said, 'I've gleaned it all from publicly available documents because I haven't yet been allowed to conduct meaningful discovery.' And he asked, 'Do you have any doubt that there is some value to chemical and biological warfare weapons?' And I said, 'I have nothing but doubt that there is value to chemical and biological warfare weapons that are deployed in such a way as to weaponize mosquitoes with things like anthrax, which would enable them to bite children along with soldiers.' There was a complete disconnect with that judge, and the case got thrown out."

Mike even called in the well-known environmental activist Erin Brockovich to lead a sit-in at the courthouse, but it did no good. Yes, that Erin Brockovich, played in the movie by Julia Roberts. Environmental activists are good roles for actors who want to get nominated for awards. But that doesn't make the job any easier for the real people on the front lines.

The case made it all the way to the Supreme Court in 2018, where it was also denied.[9]

* * *

Mike Hugo thinks the reason I haven't been able to get my case inside a courtroom is similar to why those claimants from Fort Detrick have not been able to get their claims heard. "Things happened in Judy's case which suggested to me that they couldn't be happening unless there was somebody very high up in the chain of command somewhere pulling the strings to obfuscate what this case was really about," he told Kent.

And what is my case about?

I think it's about the use of animal tissues to grow viruses for vaccines, or to use in the development of other biological products, transferring those animal viruses into human beings where they are causing a cascade of human diseases, from autism to ME/CFS, cancer, and the new diseases of aging like Alzheimer's disease when the functions of the immune system begin to falter.

We thought we knew what we were doing in those labs, but we didn't.

All of humanity is at risk, especially the lab workers.

* * *

Has Mike Hugo paid the price for his advocacy?

The evidence suggests he has.

Let's go down the list of cases he's handled. He's gone after corporate polluters, vaccine makers, pharmaceutical companies, and Fort Detrick. He's also sued the US government for something called Project Columbia, where our government sprayed glyphosate on opium fields in Columbia, only to have some of the material drift into Ecuador and destroy farms and people's health in that country.

In my case, he found himself on the opposite side of former US Senate Majority Leader Harry Reid, because my former boss, Harvey Whittemore, was his biggest donor. (It was at least some measure of justice that Harvey Whittemore was eventually charged and convicted of federal election crimes related to Senator Reid and sentenced to fourteen months in prison.)

So, it probably wasn't just handling my case that got Mike into trouble. He'd been getting himself into trouble long before he'd met me.

But my case probably didn't help.

In the last years of his life, Mike's father needed money. The recession of 2008 had hit, and after his mother died in 1983, he married a woman who blew through two and a half million dollars of his father's money. For the last few years of his life, whenever Mike's father needed money, he'd call Mike, and Mike would send him a check. By the time Mike's father died, the amount Mike had given to his father was about a hundred and eighty thousand dollars. The two of them had treated the money as a loan, and Mike's father even signed over a few of his life insurance policies to him.

However, when Mike's father died, there was a problem with Mike's sister. She'd been estranged from their father since their mother died. She decided to sue Mike and report him to the bar, claiming there'd never been any such financial agreement between Mike and their father. Yes, Mike should have had a written note with his father, but he didn't.

Mike hired an attorney to defend him, and the attorney thought the case would take two years and probably cost a hundred and fifty thousand to defend. And besides, since there was no written note, they'd likely lose. The attorney thought the worst the bar would do was write and publish a letter of censure of Mike, embarrassing him in front of the legal community, or maybe a suspension for a year.

But the bar came back saying they wanted to disbar him. Mike's wife was in the middle of her reelection campaign for a local office, and Mike said he'd accept the disbarment if it came after his wife's election. The bar agreed, his wife won reelection to her office, and Mike accepted disbarment. He was getting close to retirement age, anyway.

Mike had some friends high up in state government and thought he'd do a little investigating as to why the bar had been so severe with him. A highly placed elected state official told Mike, "If you were writing wills or recording titles and deeds, or defending petty larceny cases, you wouldn't be in this trouble. But the bar knows you're a high-profile lawyer with cases across the country, and they're using you as an example. You've taken on the high and mighty. And they want to show people you don't get to take on the high and mighty unless you're squeaky clean."

* * *

Is that how it ends?

The bad guys keep me out of a courtroom and make it so Mike can never handle another legal case?

I think life has more twists and turns of fate than we might imagine.

Mike may have had to surrender his law license, but Harvard University came calling for his expertise. He's now giving seminars on the opioid epidemic, urging treatment rather than punishment for addicts.

Mike's wife continues her political career.

And Mike's daughter, Carly Hugo, is making quite a name for herself in the film industry. She recently had her tenth film premiere at the Sundance Film Festival. Her latest film, *Share*, an eleven-minute short, won three awards at the 2019 Sundance Film Festival. The screenwriter won the Waldo Salt Screenwriting Award, the actress won the US Dramatic Special Jury Award for Achievement in Acting, and Carly and her coproducer won the 2019 Sundance Institute/Amazon Studios Producer Award for their body of work over the past ten years.

It's currently being turned into a feature film for HBO.

She's also working on a legal drama about the explosion at an oil refinery in Texas City, Texas, hoping to make it the next *A Civil Action* or *Erin Brockovich*.

What's the fate of those who fight the darkness?

Maybe it's brighter than we think.

We just need to keep fighting.

CHAPTER FIVE

Is the Government a Friend or Foe?

The government was alarmed by our research.

And after they screwed up the last serious retrovirus, HIV, and destroyed public confidence, they didn't want to make the same mistake with XMRV.

And how badly did the government screw up the HIV-AIDS epidemic?

At first, they said it was a disease that could only be contracted by promiscuous gay men, prostitutes, and intravenous drug users. You know, the original "deplorables."

But if you were a child or straight, you didn't have to worry.

Let those other people deal with the disease their lifestyle has brought upon them.

Then a child, Ryan White, got HIV-AIDS from a blood transfusion and suddenly every parent in America was thinking about that all-too-common trip to the hospital they take when their kids do something stupid like trying to jump off the roof or skateboard down that big hill.

And what about Arthur Ashe, the gentlemanly African American tennis star and Wimbledon champion, who went in for heart surgery and came out with an HIV infection, later dying of the disease?

The CDC had made terrible mistakes during the HIV-AIDS epidemic, gyrating from one extreme to the other. They say truth is the first casualty of war, and that was no less true in the fight against HIV-AIDS.

I isolated infectious HIV from saliva at the time people were being told oral sex was "safe sex." If my research had been published during the early days of the epidemic, I'm sure it would've been used to justify shunning those with the disease. But I was just a lowly lab technician at a time when technicians were not permitted to coauthor scientific papers.

As it was, my thesis research was published in the later days, as many were trying to make amends for their earlier mistakes, and it didn't fit with the narrative. My maddening frustration stems from the fact that my job as a scientist is to find out the truth and publish it.

And once we know what the problem is, we get to work on finding an answer.

How bad were things with HIV and the blood supply?

Let me do a little review for you.

* * *

A 2003 article in the *New York Times* stated it in very clear terms. Bayer had sold millions of dollars of a blood-clotting medicine to hemophiliacs in Asia and Latin America in the mid-1980s contaminated with the AIDS virus, while selling a safer version in the West:

> "These are the most incriminating internal pharmaceutical documents I have ever seen," said Dr. Sidney M. Wolfe, who as director of the Public Citizen Health Research Group has been investigating the industry's practices for three decades. . .
>
> In the United States, AIDS was passed on to thousands of hemophiliacs, many of whom died, in one of the worst drug-related medical disasters in history. While admitting no wrongdoing, Bayer and three other companies that made the concentrate have paid hemophiliacs about $600 million to settle more than 15 years of lawsuits accusing them of making a dangerous product.[1]

Think about the timeline from the act until there had been some public disclosure. Nearly two decades. That is not swift justice by anybody's reckoning.

And as the article points out, there had been fifteen years of lawsuits and somewhere around six hundred million dollars paid to victims.

Having become more familiar with corporate tactics over the past few years, we can only assume the reason we don't have a more accurate number for the amount paid out to victims is because of nondisclosure agreements, which are common in these kinds of cases.

In England it took more than three decades for the British government to issue an apology for failing to protect the blood supply in the mid-1980s.

It was only in September of 2018 that the British government spoke openly about their failure to act and conduct a public investigation of the issue as reported by the BBC:

> The government has apologized for the infected blood scandal at the public inquiry into how thousands of people became infected with HIV and hepatitis.
>
> The government's legal team said it was clear "things happened that should not have happened."
>
> It has been called the worst NHS treatment disaster.
>
> Speaking on behalf of the Department of Health and Social Care in England and its predecessor which covered the whole of the UK, Eleanor Grey QC said: "We are sorry. This happened when it should not have been done."[2]

It didn't have to be this way. Japan handled these cases much differently from how the United States or England did. They settled their HIV-tainted blood cases in 1996, as reported in the *New York Times*.

> Five drug companies and Japan's Minister of Health and Welfare agreed today to a proposed settlement with hemophiliacs who were infected with the AIDS virus through contaminated blood-clotting products, setting the stage for the end of seven years of bitter litigation.
>
> The case has roiled Japan because it seemed to show that the Government was more interested in shielding Japanese drug companies from foreign competition than in protecting public health . . .
>
> Today, several top executives of the Green Cross Corporation, one of the drug manufacturers, knelt on the floor in apology to a delegation of victims at the company's Osaka headquarters. As the mother of one victim loudly berated them, the executives bowed until their heads touched the ground.[3]

I am haunted by the facts of the Japanese case. Maybe it's only when we look at other countries that we can more clearly see the pattern of corruption in our own.

In any large-scale public health effort, it's almost guaranteed that government and private industry will share responsibility. We cannot independently verify the safety of medications. We expect there to be an extra level of protection from the government.

That was true in Japan, it was true in England, and it's true in the United States.

The classic book on the HIV-AIDS epidemic is Randy Shilts's *And the Band Played On: People, Politics and the AIDS Epidemic*, which was also turned into an Emmy Award-winning HBO film. While placing sizable portions of blame among many groups, from politicians who didn't want to talk about gays to leaders of the gay community who didn't want to implement safe sex practices because they viewed them as an infringement on their newly won sexual freedoms, Shilts also placed a good deal of responsibility on those who oversaw the nation's blood supply. In a review of Shilts's book for the *New York Times,* the paper laid out the issue of the blood supply catastrophe.

> The nation's blood supply industry at first resisted suggestions that the AIDS virus could be transmitted through blood transfusions, belittled the initial evidence that such transmission was occurring and refused to implement crude testing procedures to screen out infected blood. This, Mr. Shilts says, was largely because they did not want to shake public confidence in the blood supply, lose an important supply of blood from gay donors or pay for costly testing. Only when evidence became overwhelming and a better screening test was available did most blood banks take effective action.[4]

Even all these years later it's difficult for me to read such sections without having a visceral, emotional response. I was there, working for government science at the time these beautiful young men were dying. I talked to them, looked into their eyes, and some were my friends. I held the hand of several as they died and will forever miss their company.

The AIDS corruption and lack of humanity on the part of so many groups were eventually revealed for the entire world to see. It wasn't just about the corporations. The governments failed, as well.

They did not want to be embarrassed again with XMRV.

* * *

One of the biggest myths in public health is that articles submitted to journals are "confidential" until release and the scientific establishment only learns of the newest findings at the same time as the public.

The truth is that good ol' boy scientists at the top gossip like a bunch of old women if something interesting shows up for review at the journals.

We submitted our paper on XMRV and ME/CFS on May 6, 2009, to the prestigious journal *Science,* and even though the paper would not be published until October 8, 2009, in the summer of 2009 Frank and I found ourselves along with several other scientists at an "Invitation Only" meeting to privately discuss the findings of our work. If findings are supposed to be "embargoed" until publication, how did all of this happen?

The workshop of July 22, 2009, was titled "Public Health Implications of XMRV Infection, Center for Cancer Research (CCR), Center of Excellence in HIV/AIDS & Cancer Virology."

This is the abstract provided to the participants prior to our day-long meeting:

Introduction—In 2006, the human retrovirus XMRV (xenotropic murine leukemia virus-related virus) was identified and reported to be associated with certain cases of prostate cancer. Although the public health implications of this finding were not immediately clear, a series of presentations at the most recent Cold Spring Harbor Laboratory meeting on retroviruses provided additional support for this linkage and suggested that the number of individuals infected with XMRV is significant enough to be a cause for public concern. In view of these developments, it was deemed appropriate for NCI (National Cancer Institute) to convene a small group of intramural and extramural scientists and clinicians with expertise in this area to provide the NCI leadership with recommendations on future directions.[5]

I just want to point out that nobody from my team presented at the Cold Springs Harbor Laboratory meeting on retroviruses. As so often happens in science, that meant multiple groups, using similar new technology and new understandings, were being led to the same conclusion. Their work was preliminary and alarming, but here we were showing up with a patient population, isolation of the virus, electron micrograph pictures of the virus, and multiple confirming assays. Frank and I were the most visible faces for this discovery, but only because we had done the most work and made the most important discoveries for a quarter of a century. Who else should have been the lead on this issue?

I think it's important to highlight the two organizers of this meeting, since it shows their involvement in this issue prior the publication of our paper to the world. They were Dr. Stuart Le Grice, Head of the HIV Drug Resistance Program; and Dr. John Coffin, a professor at Tufts University, who also held an advisory position at the National Cancer Institute.

Even from the start, I believe these two individuals were "stage-managing" the potential public relations nightmare that was bearing down on the research community. It's my opinion that these two men appointed themselves guardians of the issue, so that XMRV would not get out of hand and cause the people to lose confidence, as had happened with HIV-AIDS.

Also in attendance at the meeting were two researchers from Columbia University, two from the Cleveland Clinic, two from the Fred Hutchinson Cancer Research at the University of Washington, and ten researchers from the National Cancer Institute.

All of this was because of a paper that hadn't even been published!

One of the most important action items from the July 22, 2009, meeting was that the Center for Excellence for HIV/AIDS research, in collaboration with the National Cancer Institute, authorized eight hundred thousand dollars to fund a reagent study to develop a quick and economical PCR (polymerase chain reaction) test that would cost less than a dollar and could be used to test blood samples prior to transfusion. They wanted to avoid the Ryan White and Arthur Ashe scenarios, which had scandalized the scientific community during the HIV-AIDS epidemic. We were fortunate that Frank's wife, Dr. Sandra Ruscetti, also a longtime researcher at the National Cancer Institute, was an expert on murine leukemia viruses (MLVs) and had a library of more than one hundred antibodies she'd developed over the years from this family of viruses.

That summer of 2009 things seemed to be going well with XMRV and its detection. I received emails from Dr. Ilya Singh of the University of Utah, a participant at the July 22, 2009, meeting, indicating that she was finding the correct positives in the blinded samples we were sending her.

By the time our research was published in *Science* in October of 2009, the government research community was hard at work on XMRV, trying to get right what they had gotten wrong in HIV-AIDS. In order to ensure we also had the highest level of political support for our general effort with this virus, my boss at the Whittemore-Peterson Institute arranged for Frank Ruscetti to meet with the majority leader of the United States Senate, Harry Reid.

After the meeting, Frank received a letter back from Senator Reid on his Senate stationery. The scientific details of what we were attempting would have been beyond the understanding of most politicians, and Frank recalled the senator wanted most to know if this new discovery was "solid." Frank assured him it was, based on his decades of work in the field of retrovirology, and it appeared to convince the senator, because he penned what Frank and I often referred to as the "don't mess with us" letter.

November 17, 2009

Dr. Frank Ruscetti
Head, Leukocyte Biology Section
Senior Investigator
National Cancer Institute
Laboratory of Experimental immunology
The National Cancer Institute
Building 567, Room 251
Frederick, Maryland 21702

Dear Dr. Ruscetti:

Thank you for taking the time out of your busy schedule to meet with me recently.

I appreciated the opportunity to learn more about the Whittemore-Peterson Institute's breakthrough discovery. I look forward to continuing to work with you to ensure that work is being done at a federal level to support the advancement of this important discovery.

If I can be of any assistance to you in the future, please do not hesitate to contact me.

You have my best wishes.
Sincerely,
Harry Reid
United States Senator[6]

Yes, it was a friendly little letter, but everybody knew what it meant. The majority leader of the United States Senate supported our work, and with a president of the same party in the Oval Office, Barack Obama, we felt confident we had the weight of the entire federal government behind us.

Boy, were we wrong.

* * *

Does our work with HIV and XMRV have any bearing on Alzheimer's disease? Is the explosion of Alzheimer's disease likely to be the next big public health scandal?

Consider the following article, which was published in *Medical News Today* in November of 2018 with the intriguing title "Alzheimer's May Soon Be Treated with HIV Drugs":

> New research finds that an HIV enzyme plays a crucial role in driving Alzheimer's-related brain pathology by altering the APP gene. The findings warrant "immediate clinical evaluation of HIV antiretroviral therapies in people with Alzheimer's disease," say the authors of the study . . .
>
> Currently, 5.7 million people in the U.S. are living with the condition, and the Centers for Disease Control and Prevention (CDC) predict the burden of the disease will double by 2060 . . .
>
> The APP gene encodes a protein called amyloid precursor protein found in the brain and spinal cord, among other tissues and organs. While the exact role of the APP protein is still unknown, scientists have found links between mutations in this gene and the risk of early-onset Alzheimer's cases . . .
>
> They found that the APP gene breeds new genetic variations within neurons through a process of genetic recombination. Specifically, the process requires reverse transcriptase, which is the same enzyme found in HIV . . .
>
> The researchers report that 100 percent of the brain samples that had the neurodegenerative condition also had a disproportionally high number of different APP genetic variations compared with healthy brains.[7]

I'm sure most of you followed that line of logic, but let me make sure it's completely clear. They say the best liars are those who stick close to the truth, then at the critical moment omit a crucial detail, or change the story. In this instance, they're keeping you from realizing an important truth.

The brains of those with Alzheimer's have a different genetic makeup from those who don't have the disease. If you haven't kept up with the field of genetics lately, you might simply think you have what you have. It's your genetic destiny, right?

Except for the fact that retroviruses mess everything up.

They are RNA viruses, so in order to integrate into a DNA-based organism (you!), they need a certain enzyme. That enzyme is called reverse transcriptase (RT).

The main place we ever see RT is in the presence of a retrovirus. Approximately 8 percent of the human genome is made up of endogenous retroviruses (ERVs) that have been integrated into our DNA. These ERVs can be activated to express themselves and cause damage, but for the most part they are kept silent by methylation. In addition, they are usually not replication-competent or transmissible. Thus, even if you do have many silenced ERVs in your genome, there will likely be no reverse transcriptase activity. Measuring the expression of reverse transcriptase activity is how Frank and Bernie Poiesz succeeded in isolating the first disease-associated human retrovirus, HTLV-1. I remember Frank telling me the story of watching the Geiger counter as they measured the incorporation of a radio-active tag showing RT activity as they tried to isolate HTLV-1. Usually the background counts are less than one hundred counts per minute (cpm). Their hearts raced one fateful night when they saw the counter shot up to one thousand cpm, showing they were on the right track.

But the article doesn't tell you that the rationale for using HIV drugs in Alzheimer's disease is that the overexpression of RT would indicate the activity of either ERVs or other retroviruses.

They simply say, hey, why don't we use those HIV drugs that inhibit RT? We might be able to treat Alzheimer's disease and improve the lives of millions.

Could acquired retroviruses, perhaps from animal tissues in vaccines, be expressing RT and recombining with ERVs? They don't want you to understand that. Because if you did, you'd start asking inconvenient questions.

I wonder how long it will be until the scientists who reported these findings will find themselves under attack.

CHAPTER SIX

The Blood Working Group and the Cerus Boondoggle

After the July 22, 2009, meeting at the National Cancer Institute, it was as if our research now had the official seal of approval, even though technically we were still flying under the radar, since our findings wouldn't be published until October.

Among the first calls Frank and I received after the paper was published on October 8, 2009, was from Michael Busch of Blood Systems Research Institute in San Francisco and Simone Glynn, who was head of the National Heart, Lung, and Blood Institute.

The Blood Systems Research Institute is a curious organization. It lists as its grantors the National Institutes of Health, the US Food and Drug Administration, and the US Army Medical Research and Material Command. It's physically housed within the Blood Centers of the Pacific in San Francisco. Its collaborators include the University of California, San Francisco; Oakland Children's Hospital; the US Department of Veteran Affairs; the American Red Cross; and the South African National Blood Service. It also has numerous corporate collaborations, including Cerus, Roche, and Abbott Laboratories.

The National Heart, Lung, and Blood Institute is a little less opaque. It's simply the third largest division of the National Institutes of Health and is situated in Bethesda, Maryland.

The plan was relatively simple to assess the threat to the blood supply and how to solve any problems we discovered. We'd do it in three phases.

The first was to design primers that would detect XMRV with an accurate, fast, and inexpensive PCR test. In addition, we were to develop a serology test to detect antibodies, the kind of test that had discovered Magic Johnson's HIV infection two decades earlier.

The second phase was to verify the accuracy of the tests we developed in natural samples drawn from known positive patients from the original study.

The third was to do a blinded trial with the diagnostic tests to determine their sensitivity and specificity. We did not want to misdiagnose an infected person as not infected and vice versa.

The antibodies we used in our serology test as published in our *Science* paper were excellent for the detection of XMRV and remain so to this day. It was the only test that detected all XMRV family members accurately. It should have been enough, if we were really trying to determine if there had been exposure to XMRV.

By way of comparison, Magic Johnson never had infectious HIV isolated from his system. His blood test showed antibodies to HIV, meaning he'd been exposed to the retrovirus and could be expected to develop AIDS and die from it.

That's why getting him started on the antiretrovirals so early was important. They silenced expression of the virus, which is how the immune system gets damaged.

But I don't believe those in charge were simply interested in having a reliable test. There were too many other factors in play.

* * *

XMRV was poised to be much bigger than HIV.

When Magic Johnson was diagnosed in 1991, it was estimated that one million Americans were infected with HIV. Our research showed 3.75 percent of the population was carrying this virus. Columbia University professor Ian Lipkin's multicenter study and his later study with Dr. Montoya of Stanford University showed about 6 percent of the population was carrying the virus. Lasker award-winning researcher Harvey Alter and his colleague Shyh-Ching Lo found 6.6 percent of the healthy controls to be positive for XMRV and related members. We had been on the conservative side in estimating the size of this epidemic. If you take these numbers as a range, there are somewhere between ten and twenty million Americans carrying some variant of the XMRV family of viruses.

I believe the lesson of HIV-AIDS is that when there's a threat to a single person or group, it's a threat to all of us.

The mistake we made in HIV-AIDS was to look from a position of moral superiority on the gay and drug-using population, instead of seeing the problem as one for the entire human family. What affects one affects us all. That's not simply a spiritual perspective, but one that is grounded in solid science. Spraying pesticides on the plants to kill insects is also likely to harm a lot of other living things, including humans.

Some may fault us for our approach, but it made no sense that XMRVs would be solely linked with ME/CFS and prostate cancer. That was not what the data from three years of family studies suggested, and it was not what we had learned about retroviruses. The virus had first been found in prostate cancer tissue. Last I checked, women don't have prostates. That's why we were so focused on the family studies and seeing the patterns of disease in those groups. Were we surprised to find autism in the children of mothers affected by ME/CFS?

We would not have predicted these results at the outset. But once the data suggested it, the pathology made a good deal of sense to us.

For God's sake, for decades we'd been advising pregnant women infected with HIV to put their children immediately on antiretroviral drugs prior to any immunization for fear that a vaccine might trigger full-blown AIDS. The virus liked to hide out in the monocytes, the B and T cells of the immune systems, exactly those cells a vaccination would stimulate.

This wasn't rocket science.

All we were saying was if this time bomb was already in a good percentage of the population, we didn't want to be setting off an explosion of neuroimmune disease and cancer with a vaccination.

In 2010, it was estimated that the cost to test a patient for HIV was about twenty-two dollars. At that time, it was estimated that approximately thirty-four million people were infected with HIV around the world. Every one of them required a blood test. That meant hundreds of millions of dollars for whoever owned the right to the test.

Let's take the lower figure that approximately 4 percent of the United States was infected with XMRV in 2010, shortly after the publication of our research. That's twelve million Americans. If we looked at that in terms of the world population at that time, it meant somewhere around 275 million people around the globe were infected with XMRV.

An effective blood test for XMRV wouldn't be worth hundreds of millions of dollars.

It would be worth billions.

You probably won't be surprised to learn that Frank and I were the sole inventors of XMRV along with our research institutions in a patent application dated April 6, 2010, for "Strains of Xenotropic Murine Leukemia-Related Virus and Methods for Detection Thereof," and received the patent application number 20110311484.

The abstract to our application stated:

> Provided are novel strains of Xenotropic Murine Leukemia Virus-Related Virus (XMRV), or polynucleotides or polypeptides thereof. Identified herein are nucleic acid changes or amino acid changes identified in XMRV strains isolated from subjects. Also provided are methods of detecting such XMRV strains based at least in part on the identified nucleic acid changes or amino acid changes.[1]

A couple of things are important for you to notice. As you can tell, we distinctly refer to there being multiple strains of XMRV, and we had isolated these strains from actual patients. From the beginning, our data in the 2009 *Science* paper suggested there were multiple strains. We also isolated the variant strains from actual patients, not like others who used VP62, a molecular viral clone as a stand-in for the actual virus.

* * *

After our research was published in October of 2009, I was deluged with phone calls.

Many were from patients with ME/CFS who wanted to know what it meant for their disease, some were from clinicians who felt we might have finally found the last piece of the puzzle for many of their patients, and one was from Cerus, a company in Concord, California.

As I recall, it was one of the vice presidents of the company, Dr. Lily Lin, who spoke on the phone with me. She was impressed with the paper and said it dovetailed with some studies they'd been doing regarding decontaminating retroviral threats to the blood supply. And more important, she believed her company had a technology, the INTERCEPT system, that could safely neutralize any RNA viruses, including retroviruses, which might be in a blood sample.

It was not lost on us given our experience with HIV, but if you could easily and reliably decontaminate any blood donations that came in, you probably didn't need a test. The value of a test might be lower if you could simply choose to decontaminate all blood.

I gave Lily a presentation in a coffee shop near the company head-quarters, showing the data of the viral proteins expressed and antibodies made to them and showing we had more than simply an incomplete genetic sequence. It brought me back to what Robert Silverman of the Cleveland Clinic had told me when we first started sharing findings. "I hope you can get an immune response," he said. "Because we've done thousands of sam-ples and they're all negative." Of course, they were negative, because he was using his Frankenstein stitched-together molecular clone of the virus, named VP62, which has never existed in an actual human being.

The Abbott blood test didn't work because it was not detecting the sequences that had been isolated from a human being or recognize the pos-sibility of a conformational epitope. If you have the wrong sequence, the corresponding protein won't fold into the shape detected by our original serology assay. (Yes, we will talk about why you never use an infectious molecular clone when you're studying viruses in the next chapter.)

So, what we did with Cerus was more or less exactly what I'd done at Upjohn thirty years earlier. We simply sent samples of blood from individu-als infected with XMRVs. They'd use their INTERCEPT technology on it, then send it back to us to see if the blood was now free of infectious XMRV. The Cerus technology worked beautifully.

At the time I was fired from the Whittemore-Peterson Institute, my lab was working on a study with Cerus that was looking at patients from Oakland Children's Hospital who'd received multiple blood transfusions. If retroviruses were present in the blood supply, it meant we'd find a higher rate of XMRVs in the transfused children than in any of our con-trol groups. I couldn't finish the study because of being fired and the data removed from my office. But the preliminary set of data from those chil-dren showed an infection rate of around 8 percent. I have never seen the published results of that study and suspect it was simply filed in a drawer to be forgotten or, worse yet, destroyed. Eight percent was significantly higher than the 3.75 percent we'd found in our original science paper, or the 6 percent Lipkin found in his multicenter study, or the 6.6 percent that Harvey Alter and Shyh-Ching Lo found in their confirmation study of our work.

On March 29, 2011, I presented our findings on XMRV at the New York Academy of Sciences. At the end, I noted my five main points as being:

1. Data suggest there are different strains of gamma-retroviruses that can infect humans.
2. Assays that capture the variation of these viruses in the blood supply are the best, i.e., serology and transmission.
3. Cerus Technologies can inactivate strains of XMRV/HGRVs in blood components.
4. New disease associations include leukemia, lymphoma, and the platelet/megakaryocyte blood disorder ITP.
5. We need more full-length sequencing.

I think it's clear that while we had little trouble discussing the public health risk suggested by the data, we also had little hesitation suggesting solutions that were supported by our collaboration. Cerus had a great product that would deal with the threat to the blood supply.

The only problem, I realized much later, was that the government didn't want to admit there had been a threat to the blood supply. Our 8 percent finding of XMRVs in children who'd received multiple transfusions through Oakland Children's Hospital hinted at enormous legal liability for the blood banks. That applied equally to the government agencies, which had verified the safety of those blood banks, and the blood supply in general, to the public.

* * *

"Agency heads are scared to death of how the patient population will react if XMRV works out." This is a direct quote from the well-known ME/CFS patient website Phoenix Rising. It was alleged by a forum member to have been said on September 11, 2010, by Suzanne Vernon, the scientific director of the Chronic Fatigue Immune Dysfunction Syndrome Association of America (CFIDS). This allegedly happened in a discussion between the forum member, Suzanne Vernon, and advocate Cort Johnson during a break at the 2010 OFFER Utah Patient Education Conference in the lobby of the Salt Lake City downtown Hilton.

The forum member, identified as "CBS," a senior member of the Phoenix Rising forum with more than 1,400 posts at the time, added, "I've been struggling [with] what I ought to do with this for almost six months. Suzanne Vernon said this during a conversation she was having with me and Cort. She just sort of interjected it. No real need, nor was there much of a segue. She said that it should not be repeated. Yet I wondered why on Earth she would say something like that to someone she had just met."[2]

I include this account to reinforce my feelings that people with whom I had no connection believed that our federal government did not want XMRV to work out.

And why not?

Well, as I've said before, it would be a huge liability for the United States government.

While I couldn't personally attest to Suzanne Vernon making those comments in September of 2010, less than a year after our publication in *Science*, I was also observing some unusual behavior among government and corporate researchers.

Why might Suzanne Vernon have been reporting on September 11, 2010, that agency heads were worried about XMRV working out? Maybe because on September 7 and 8, 2010, the First International Workshop on XMRV took place, and there was a good deal of research that supported our findings.

Maureen Hanson of Cornell University submitted an abstract titled "XMRV in Chronic Fatigue Syndrome: A Pilot Study," which looked at ten subjects with severe CFS, ten who had "recovered," and ten controls. Dr. B. P. Danielson of Baylor University contributed "XMRV Infection of Prostate Cancer Patients from the Southern United States and Analysis of Possible Correlates of Infection." From Emory University came more evidence of XMRVs in prostate cancer, as documented in their abstract "Variant XMRVs in Clinical Prostate Cancer." The Cleveland Clinic and Robert Silverman submitted two abstracts, "Presence of XMRV RNA in Urine of Prostate Cancer Patients" and "XMRV Infection Induces Host Genes that Regulate Inflammation and Cellular Physiology." Paul Cheney, who would go onto work with Jeff Bradstreet up until his mysterious death, submitted a positive paper titled "XMRV Detection in a National Practice Specializing in Chronic Fatigue Syndrome (CFS)."

A troubling result came from a paper submitted that was a collaboration between Emory University, the Cleveland Clinic, and Abbott Laboratories. Their abstract, "XMRV Induces a Chronic Replicative Infection in Rhesus Macaques Tissue But Not in Blood," told a horrifying tale, both for sufferers and researchers attempting to make sense of it all. They found the virus disappeared relatively quickly from the blood but would replicate to detectable levels upon any immune stimulation, such as from a vaccine.

In other words, a vaccination could awaken a sleeping monster.

This was the very scenario I had publicly suggested shortly after publication of our work.

Lasker Award winner for his discovery of the hepatitis C virus, Dr. Harvey Alter, was finally able to talk about his long-delayed paper that had finally been published in the *Proceedings of the National Academy of Sciences*. He wrote:

> We found MLV-like *gag* gene sequences in 32 of 37 patients (86.5%) compared with only 3 of 44 (6.8%) healthy volunteer blood donors. . . . No evidence of contamination with mouse DNA was detected in the PCR assay system or any of the clinical samples. Seven of 8 *gag*-positive patients were again positive in a sample obtained nearly 15 years later.[3]

I thought the most controversial abstract would be the one I'd worked on with Frank Ruscetti with the title "Detection of Infectious XMRV in the Peripheral Blood of Children." We wrote:

> [A]n understanding of XMRV infection rate in children may be particularly helpful, given that 1 in 100 children in the US are diagnosed with neuro-immune disorders, including Autism Spectrum Disorder (ASD) and that CFS and childhood neuroimmune disorders share common clinical features including immune dysregulation, increased expression of pro-inflammatory cytokines and chemokines, and chronic active microbial infections.
>
> XMRV was detected in 55% of 66 cases of familial groups from 11 states. Sequencing of PCR products of *gag* and *env* confirmed XMRV. The age range of the infected children was 2–18. 17 of the children (including the identical twins) were positive for XMRV (58%) and 20 of the 37 parents (54%) were positive for XMRV. 14 of 17 autistic children were positive for XMRV (82%).[4]

That should have been the bombshell.

Not only were we saying we'd found a pathogen responsible for the cascade of immune dysfunction resulting in ME/CFS, but that this same retrovirus family may have also given birth to the autism epidemic. The terrible lesson of the HIV-AIDS epidemic is that there are no "deplorables" among us. We are all connected, and if any one of us suffers, that pain will eventually reach everybody.

But it wasn't children, suffering families, or autism that attracted the attention of Dr. Francis Collins, head of the National Institutes of Health.

Collins cared about the blood supply.

* * *

Collins attended the opening of the session, held in the Masur Auditorium in the main building of the National Institutes of Health, located in Bethesda, Maryland. He listened for a short while, then left. He reappeared on the second day to listen to my talk. We'd put together an abstract titled "Detection of Infectious XMRV in the Peripheral Blood of Chronic Fatigue Syndrome Patients in the United Kingdom," which attracted serious attention.

England was one of the epicenters of ME/CFS, and I felt a strong sense of responsibility toward them for the support they'd always given our efforts. The investigation we'd put together in the United Kingdom had been dramatic, involving nearly fifty patients, arranging car rides for people to blood draw centers in churches, or even a few of the patients receiving visits from phlebotomists at their homes to get a sample.

We thought the United Kingdom study was a tour de force of scientific research, especially since the samples were not drawn or processed in a lab. They were processed in churches and homes, essentially eliminating any possibility of contamination. Our findings were similar to what we'd previously reported among sufferers of the disease, and the controls were also in line with previous findings of about 4 percent in healthy controls.

I felt as if everybody in the auditorium turned to look at Francis Collins when he raised a hand to ask a question.

"Yes?" I said, my heart pounding.

"Where did you get the 4 percent positive controls?" Collins asked.

"The blood supply in the UK," I explained, noting we'd also obtained fifty or sixty additional positive controls, age and sex matched to the UK ME/CFS patients.

I think that was the moment Collins decided he needed to call Dr. Tony Fauci, head of the National Institute of Allergy and Infectious Disease (NIAID). Finally, no longer would these suffering patients be considered crazy and denied the same type of therapies that had ended the threat of HIV-AIDS. But little did I know that Dr. Fauci was not about to let this happen. What did Fauci do? He called in Dr. Ian Lipkin to lead that validation study.

Dr. Ian Lipkin of Columbia University, the great debunker, who had participated in a similar shameful episode against a doctor I've now come to know well, Dr. Andrew Wakefield. In my estimation, Dr. Wakefield is our greatest scientific martyr, for his initial suspicions voiced in a series of case reports for *The Lancet*, of a connection between the MMR (measles-mumps-rubella) shot and the development of gastrointestinal problems and autism.

When I met my attorney Mike Hugo, he said, "This might sound a little strange, but I've got an expression for what I think has happened to you."

"What's that?" I asked.

"You've been Wakefielded."

I laughed. "I know Andy." We had talked at length about a study we had done looking for an association of XMRV with ITP, in collaboration with Dr. James Bussels of New York City. (ITP is idiopathic thrombocytopenic purpura, a disorder that can lead to easy and excessive bruising and is thought to result from low levels of platelets that help with blood clotting. ITP is a CDC-acknowledged side effect of MMR. We found evidence of ITP in 30 percent of the sixty or so patient samples we tested with our serology test. Interestingly, there was no correlation with fatigue.)

I thought speaking to the head of the National Institutes of Health about a solution to our biggest health crisis since the HIV-AIDS epidemic was the moment when science would fulfill its most important duty to the public.

Instead, Collins was setting me up with Lipkin, the same man who had Wakefielded Dr. Andrew Wakefield. And looking over it all would be Tony Fauci, head of the National Institute for Allergy and Infectious Diseases, and the boss of Kuan-Teh Jeang, editor of the journal *Retrovirology*, who would also eventually die under mysterious circumstances, just like my friend, Jeff Bradstreet.

I never realized scientific and medical research could be so dangerous to a person's health.

* * *

That first meeting with Dr. Ian Lipkin of Columbia University on November 4, 2010, was ostensibly to design a multicenter study to confirm our findings regarding XMRV and ME/CFS.

If that were so, why were there members of the Blood XMRV Scientific Working Group (called Blood Working Group) present? People like John Coffin; Simone Glynn of the National Heart, Lung, and Blood Association;

and Susanne Vernon, former veteran of the Centers for Disease Control, who only two months earlier expressed that those same agency heads were "scared to death" of the XMRV association with ME/CFS. What were they doing there? The Blood XMRV SWRG AKA "Blood Working Group" had nothing to do with the association of XMRV with ME/CFS!

I think somebody up very high had decided that XMRV and this whole issue had to go away, and they'd assembled the team to make it happen. The assassination had been planned, and I was the only one in the group who didn't know it.

If there had been no effort to control the discussion about XMRV and ME/CFS as well as a myriad of other diseases, here's how I think things would have gone.

In October of 2009, we published our findings in the journal *Science*.

In a meeting on November 10, 2009, Dr. Gary Owens of the University of Virginia told us he'd confirmed XMRV-2 in cardiac tissue. Now we've got some heart problems connected to that retrovirus family.

In a closed meeting in Zagreb, Croatia, on May 25 and 26, 2010, Dr. Harvey Alter confirmed our findings as reported in a Dutch health magazine. A direct quote from the article states:

> The FDA and the NIH have independently confirmed the XMRV findings as published in Science, October [2009]. The confirmation was issued by Dr. Harvey Alter of the NIH during a closed workshop on blood transfusion held on May 25–26 in Zagreb . . . The association with CFS is very strong, but causality not proved. XMRV and related MLVs are in the donor supply with a prevalence of 3% and 7%. We (FDA and NIH) have independently confirmed Lombardi group findings.[5]

Pretty much sounds like game over, doesn't it? You have an award-winning researcher working with a team that includes scientists from the Food and Drug Administration and the National Institutes of Health, and they confirm your work. In addition, they find somewhere between 3 percent and 7 percent of the population infected with this virus or related family members.

Time for action and getting to work solving the problem, right?

No, this is where I believe the sabotage began.

I think this is when they realized they had an unsafe blood supply, they'd harmed a lot of people, and there would be enormous liability for blood banks and government agencies when this information was revealed.

The sabotage was done by using Robert Silverman's synthetic infectious molecular clone as a reference for the virus and taking me out of the equation. In phase two of the blood working group, I'd been the one blinding the samples, so the researchers didn't know which samples were positive and which were negative.

Remember, we were supposed to come up with a quick and inexpensive PCR diagnostic test in the blood working group. And the truth was, as expected, Switzer and the CDC researchers weren't getting good results from the PCR based on the VP62 Frankenstein clone. I realized the problem was likely natural isolates of XMRV in the infected people were usually latent/silent. Only in the sickest patients could XMRV be detected in the blood. This had been our seminal work in HIV latency. You see, most retroviruses contain many cytosine/guanine pairs, CpGs in the promoters. This means the on/off switch of XMRVs turned off and the virus was silenced by the addition of methyl groups and would not be detected in a 30-second PCR test. That fact that the XMRV on switch/promoter was a highly-methylated molecule was critical. Most other viruses do not have so many CpG pairs in the on/off switch. They are easier to pull apart. This fact might be lost on the general public, but every virologist worth their degree knew what it meant. A quick-and-dirty PCR would never reveal the secrets of this virus and thus would be a silent contaminator of the blood supply.

When I provided this explanation at the September 22, 2011, Ottawa conference where I squared off against John Coffin, I had so many people come up to me and say, "That's brilliant! That explains the data perfectly."

Let me say that again.

XMRV is tightly bound with high levels of cytosine and guanine in a configuration that is difficult for a typical PCR test to detect. The typical thirty-second denaturization procedure of a typical PCR test is not enough to pull these molecules apart and reveal their secrets.

We needed to find another way to test for this virus or simply use the Cerus INTERCEPT technology to decontaminate the blood supply. But the simple fact was that the *Science* publication of September 22, 2011, from the Blood Working Group was going to be used to discount the association of XMRVs to ME/CFS, not to discuss the contamination of the blood supply, which was in fact the ONLY mission of the Blood XMRV SWRG. These "scared-to-death public officials," to use the words of Suzanne Vernon, were not going to wait more than a year for the official Lipkin debunking show.

They were going to fraudulently publish the results from the Blood Working Group to slit my throat and hope I bled to death on the floor.

* * *

How's this for a perfect storm?

My employer, the Whittemore-Peterson Institute of the University of Nevada, Reno, was funded by the real estate holdings of the Whittemore family. The recession had devastated their finances, as it had many large landowners, and they were hopeful we'd come up with a PCR test for the blood supply. That would have done wonders for their financial picture and the people from whom they'd borrowed money. It was potentially worth billions of dollars.

But there I was saying that any PCR test or serology test based on the VP62 clone was worthless, including the one their company VIPDx had been selling. Abbott had licensed that clone from Silverman and expected it to be worth billions of dollars. Was this why Silverman kept pushing VP62, even when it became clear that wasn't the sequence of XMRV, or even a useful model?

On the other side was a public health establishment, desperate for XMRV to be debunked, thus absolving them of billions of dollars of potential legal liability.

And instead of supporting the establishment cause, there I was, saying, no, XMRVs are real and are probably responsible for many health conditions and were in many cases likely to have been transmitted through the blood supply.

I always say science doesn't have a side. It's just about truth. But that doesn't consider our modern world and how it's corrupted science.

On one side were scientists who'd opened an area of inquiry, but they had the wrong sequence for the virus, and their tests were worthless.

On the other was the public health establishment, which simply wanted it all to go away.

And there was me, saying, yes, there is something here, but we don't have it yet nailed down. But when we do figure it out, it's going to be larger than anybody can imagine.

How bad did it get?

Worse than you can imagine.

* * *

I got a call from Michael Busch on August 31, 2011, telling me I had to finish final edits of the paper reporting the results of the Blood Working Group a week later (the Tuesday after Labor Day) because that was the last day if it was to be published by *Science* on September 22, and it would now be titled "Failure to Confirm XMRV/MLVs in the Blood of Patients with Chronic Fatigue Syndrome: A Multi-Laboratory Study."

I was enormously frustrated and angry.

The Blood Working Group was tasked with finding whether a fast and inexpensive PCR or serology test could be developed for the blood supply. If they had wanted to write—we had failed to develop an accurate test for the blood supply—that was fine with me. That was accurate. And I thought I could explain that failure in a way that made scientific sense.

The title suggested we'd done an association study, which we had not.

There were only fifteen patients in the Blood Working Group sample.

That was not an association study.

I was being given a single day to approve the paper as written, and as written it was simply fraud.

This was not science. It was pure politics.

Here's the email I sent back to Simone Glynn on August 31, 2011, at 8:24 p.m., PDT, of that same day. If this was pure politics, I'm sure you'll agree I'm no diplomat. But I am a fighter. I wrote to Simone in response to her request to approve the paper as written in a single day in order to have a record of what was being done:

> That's impossible.
>
> I have IRB protected data that I cannot access until the 6th. I told that to Graham yesterday and he indicated it was fine. Given the complexities and limitations of this study, many of which were not recognized at the time the (flawed) experimental design was agreed upon. To have one day to agree upon a manuscript, a holiday at that, is totally unacceptable. This is NOT good science or the appropriate process. What is the rush?
>
> Afraid of the truth??? How many of these viruses were introduced into the human population and are now threatening a lot more than the blood supply??? Because a few declared it "impossible" 40 years ago and JC [John Coffin] himself was the most vociferous??
>
> How many XMRVs??
>
> I am sending this only to Simone and Frank because I will make this rush a public relations nightmare for the entire US govt. I have integration

data and variants of many new strains!! Did those arrogant SOBs introduce these into humans and are now trying to cover it up??

And then pedigree the negatives with a cutoff so high it would not find a willing woman in a whore house?? Wonder if anyone will listen to a press conference from me?? Asking how many new recombinants from vaccines? From lab workers?? Doctors? The first ever contagious human retrovirus???? Spread like mycoplasma?? Are you kidding me???

It happened once!!! How many xenograft lines were created??? How many vaccines contained mouse tissue??

These sick people lost their entire lives and this travesty of justice will not be carried out at their expense. Not again.

If we have to write and publish online a dissenting opinion, we will and I will not coauthor any paper that misrepresents our findings. Nor will our data be included. You can simply say we all found nothing. Totally expected and we'll prove them all wrong.

Our assays may not be sensitive or reproducible given the complexity and lack of knowledge of reservoirs, etc. Nothing about these data say anything about Lombardi et al, Lo et al. Except that there are likely many strains of XMRVs and God only knows the impact on chronic disease. But nothing about this study says anything about our original discoveries.

And if this is rushed to print without a fair and balanced discussion of its limitations, I will spend every minute of my life exposing the fraud that has been perpetrated against this patient population.

Judy Mikovits.

Yes, I was heated, but I had good reason to be angry. This email generated a phone call that Labor Day from Michael Busch and Simone Glynn.

Simone said, "I have Harold Varmus [Head of the National Cancer Institute at the time] on the phone, and if you and Frank do not coauthor this paper, Frank and Sandy will be fired for fraud (for our 2009 *Science* paper) immediately and lose their pension. They will lose their entire retirement."

Yes, Simone Glynn, head of the National Heart, Lung, and Blood Institute, threatened the pensions of two senior scientists, claiming to have the support of the head of the National Cancer Institute on her side.

Mike Busch and Simone said they'd change the language of the article and the title to language acceptable to me. But they didn't, and it got published with the fraudulent title and data interpretation.

I think that paper is the greatest example of scientific misconduct I've ever witnessed in my life, and everybody who willingly put their name on it should be driven out of science.

There were only fifteen patients in the study. That's not an association study.

It was a cover-up of the contaminated blood supply.

In the article written for *Science* about my face-off with John Coffin on this issue at the September 22, 2011, Ottawa IACFS conference, and published on September 30, 2011, Jon Cohen ended the piece by saying, "Mikovits said she hopes to have full sequences of her new viruses in a couple of weeks."

I was fired the day before that article came out because I wouldn't say the diagnostic test my employers were using in their private company, VIPDx, was validated, nor allow the director of that company to say he had worked on a government grant when he had not. Once again, my honesty got me in trouble. Harvey Whittemore knew exactly which of my desk drawers contained the actual sequence of the virus. How convenient for Jon Cohen to write that article, knowing I'd never see those sequences again!

I had no lab, I was prevented from getting a new job with the grants, which had been awarded to me as principal investigator. I was jailed, and the contents of two offices, the data of those of my research teams, and the work of my entire career were all locked down, never to be seen by me again.

Is that the way we are to do science in the future?

If a scientist uncovers troubling information that affects the lives of millions, that information will simply be locked away, while perpetrators of scientific fraud and crimes against humanity are rewarded with millions of dollars? It's almost like 1934 in Los Angeles with the sick medical staff all over again.

It's almost as if I never existed.

* * *

Cerus was the one who made out like a bandit.

Maybe I shouldn't be so disturbed by the behavior of Cerus. They were simply looking to provide a service and make money, not get in the middle of what was potentially a multibillion-dollar fight between patients and the United States Government.

The INTERCEPT Blood System Technology is truly revolutionary. As described in their 2017 Annual Report, the blood samples are:

> [P]laced in an illumination device, or illuminator, where the mixture is exposed to ultra-violet A or UVA light. If pathogens such as viruses, bacteria or parasites, as well as leukocytes, or white cells, are present in the platelet or plasma components, the energy from the UVA causes the amotosalen to bond with the nucleic acid. Since platelets and plasma do not rely on nucleic acid for therapeutic efficacy, the INTERCEPT Blood System is designed to preserve the therapeutic function of the platelet and plasma components when used in human transfusions.[6]

You see, science is up to the challenge of solving difficult problems. I also think it helps if the truth is told, but apparently, that's not required in public health.

In 2009, Cerus reported sales of $16.8 million against operating expenses of 29.2 million dollars, noting their fourth quarter sales had gone up 51 percent over the previous year, crediting the rise to growing acceptance of their product in France, Belgium, Southern Europe, and the Middle East. In September of 2010, a press release from Cerus demonstrated the success we'd had with their technology:

> The Whittemore-Peterson Institute for Neuro-Immune Disease (WPI) and Cerus Corporation presented data at today's NIH-sponsored 1st international Workshop on XMRV which demonstrates the efficacy of Cerus; INTERCEPT Blood System to inactivate XMRV and other MLV-related viruses in donated blood. Recent scientific studies have detected these human retroviruses in up to seven percent of healthy blood donor samples, indicating approximately 20 million people in the United States could unknowingly be carrying the infection of XMRV and MLV-related viruses have been linked to prostate cancer and myalgic encephalomyelitis/chronic fatigue syndrome (ME/CFS).
>
> "The genetic variability of XMRV and MLV-related viruses will make development of screening assays for the blood supply challenging," said Dr. Judy Mikovits, director of research at WPI and lead author of the study. "The INTERCEPT technology demonstrates robust inactivation of these viruses and holds promise as a potential proactive approach to mitigating the risk of XMRV/MLV-related virus transmission via transfusion."[7]

The same press release highlighted what should have been definitive confirmation of XMRV by leading government research institutions as well as the methods we used, which did NOT include VP62, the infectious molecular clone:

> In a paper published online on August 23, 2010 in the journal *Proceedings of the National Academy of Sciences* (PNAS), scientists from the National Institutes of Health and U.S. Food and Drug Administration detected the presence of a genetically diverse group of MLV-related viruses in 86 percent of CFS patient samples and in 6.8 percent of samples from healthy blood donors, leading to new concerns about the possibility of transfusion transmission. The PNAS study results are consistent with data from a 2009 study published in *Science*, which detected XMRV in 67 percent of CFS patients and 3.7 percent of healthy controls.
>
> In the study conducted by WPI and Cerus, red blood cell and platelet components were contaminated with a *natural isolate* (italics mine) of XMRV and MLV-related viruses from an ME/CFS patient. INTERCEPT-treated and control samples were evaluated in a *validated virus culture test* (italics mine), which allows sensitive detection of viral particles that are capable of reproducing. No viable virus was detected following treatment, indicating the intercept Blood System is capable of inactivating high levels (>4 logs) of the virus.[8]

Just to recap what the previous paragraphs revealed: scientists from the National Institutes of Health and the Food and Drug Administration in 2010 found XMRV-related viruses in 86 percent of patients with chronic fatigue syndrome and in 6.8 percent of healthy controls.

This was an even more robust finding than our 2009 paper showing XMRV in 67 percent of patients and in 3.7 percent of healthy controls.

Our research DID NOT focus on a PCR test, but a virus culture test, which allowed for detection of viral particles capable of reproducing and proteins and immune response serology testing.

The Cerus INTERCEPT Blood System seemed to completely eliminate the threat to the blood supply posed by XMRV, at least to the best of our detection technology.

We were the ones doing good science.

However, our good science would raise serious questions about the dangerous practices of the past, and how many were today suffering as a result of those mistakes.

* * *

In a January 7, 2019, press release, Cerus announced they'd exceeded their 2018 projection of fifty-eight million to sixty million in revenue, bringing in $60.9 million. They expect 2019 to be an even better year, estimating sales between seventy and seventy-three million dollars.

> "The revenue growth we generated in 2018 underscores the increasing demand for safer blood components. We finished 2018 strong with quarter-over-quarter and year-to-date growth in disposable kits led by French national conversion and U.S. demand," said William "Obi" Greenman, Cerus' president and chief executive officer.
>
> "Over the past few months, U.S. customer orders for INTERCEPT platelets have been increasing. With the recent FDA publication of the draft guidance document on bacterial risk control strategies for platelet collection and transfusion, we could potentially experience further acceleration in customer demand in the U.S.," continued Greenman.[9]

Perhaps it's a little too much to expect a company making sixty to seventy million dollars a year to point out a liability to the United States government of billions of dollars.

But wait, I'm a little confused. The government says there's no threat to the blood supply. So why is a company making tens of millions of dollars making the blood supply safe, when according to our leading public health authorities it's already safe? Are they just being extracareful?

I can imagine a corporate board meeting at Cerus where the executives ponder this question. On the one hand, they say to themselves, if we go along with the government's lies, we'll make a lot of money and be a successful company. On the other, if we tell the truth, we can expect within a short time to be raided by government agents on some flimsy pretext. And even if we prevail in court, the bad publicity will cause the company to fail. Which direction would the executives choose?

This is what the government wants you to believe. The blood supply is totally safe. They just want to make it safer by spending tens of millions of dollars on a completely unnecessary system. Pay no attention to what may or may not have happened in the past.

Am I being too sarcastic?

Let me give it to you straight, then.

The blood supply was and is not safe, and Cerus and the various government agencies know it.

As a private individual, I thought I could defend my scientific reputation by exercising my constitutional rights. I was wrong. I still haven't had a single day in court, and my attorney has been disbarred for giving money to his elderly father without having a loan agreement.

Still, I can't help but think that somebody out there somewhere in a position of influence, who can tip the balance toward truth and disclosure, will someday listen to the voice of their conscience and act.

I'd like to wish Cerus well, because I believe their technology is saving lives.

However, as they are part of what I consider to be a continuing criminal enterprise to cover up the truth about lab-derived animal retroviruses and microbial agents infecting the public and causing many different diseases, I must reluctantly decline.

CHAPTER SEVEN

VP62—The Clone Assassin

Biology does not necessarily follow technology.

That may be one of the most important things I write in this book, so let me emphasize it again.

Biology does not necessarily follow technology.

Let me explain what I mean by that statement.

Just because you come up with a new piece of technology, which lets you know more than you knew before, doesn't mean you now know EVERYTHING.

Scientists must have humility when approaching the unknown. The rush to judgment is a critical error many scientists made in the XMRV debate, even regardless of whether there was financial pressure for them to come to a result different from ours.

I love technology, but I'm also aware of its limitations. I'll always take the new information to deepen my understanding. But a good scientist will also realize the limitations and unanswered questions that remain.

* * *

How did all this start?

Why did people start looking at a mouse virus and its link to conditions like prostate cancer in men, ME/CFS in an overwhelmingly female population, and diseases like autism in children? And might this retrovirus family or another one much like it be linked to the new diseases of aging, like Alzheimer's?

All of that started with a new piece of technology that made an unexpected discovery of mouse-related retroviral sequences in tissues from prostate cancer tumors. Joe DeRisi is a professor at the University of California,
San Francisco, with an interest in how technology might unlock biological
processes, such as how viruses might be driving the development of cancer.

DeRisi chose a TED Talk on January 29, 2006, to discuss his techno-
logical innovation, known as the ViroChip, and what he had discovered
when Robert Silverman of the Cleveland Clinic hypothesized men with a
genetic defect resulting in an inability to degrade RNA viruses would be
more susceptible to retroviral infection. This would lead to these men having more aggressive cancers. Silverman provided DeRisi with some tissue
from prostate cancer tumors. DeRisi tested the tissue to prove or disprove
that hypothesis.

DeRisi opened his talk by asking questions of the assembled audience:

> How can we investigate this flora of viruses that surround us, and aid medi-
> cine? How can we turn our cumulative knowledge of virology into a simple,
> hand-held, single diagnostic assay? I want to turn everything we know right
> now about detecting viruses and the spectrum of viruses that are out there
> into, let's say, a small chip.
>
> When we started thinking about this project—how we could make
> a single diagnostic assay to screen for all pathogens simultaneously—well,
> there's some problems with this idea. First of all, viruses are pretty complex,
> but they're also evolving fast . . .[1]

Everything about DeRisi's opening was perfect. What scientist doesn't want
to streamline what we know about viruses into a simple test that can be used
to quickly identify a potential pathogen impacting health?

I know I do.

And DeRisi also noted that viruses are complex and evolve quickly. The
majority of DeRisi's talk concerned the development of his chip and the
complexity of viruses, but he saved the bombshell for the end:

> I'm going to tell you one thing in the last two minutes that's unpublished.
> It's coming out tomorrow. And it's an interesting case of how you might use
> this chip to find something new and open a new door. Prostate cancer. I don't
> need to give you many statistics about prostate cancer. Most of you already
> know it: third leading cause of cancer deaths in the U.S. Lots of risk factors,
> but there is a genetic predisposition to prostate cancer.

For maybe about 10 percent of prostate cancer, there are folks that are predisposed to it. And the first gene that was mapped in association studies for this, early onset prostate cancer, was this gene called RNASE-L. What is this? It's an anti-viral defense enzyme. So, we're sitting around and thinking, "Why would men who have this mutation—a defect in an antiviral defense system—get prostate cancer? It doesn't make sense. Unless, maybe, there's a virus?"

So, we put tumors—and now we have over 100 tumors—on our array. And we know who's got defects in RNASE-L and who doesn't. And I'm showing you the signal from the chip here. And I'm showing you the block of retroviral oligos [oligonucleotides—short RNA or DNA molecules]. And what I'm telling you from the signal, is that men who have a mutation in this antiviral defense enzyme, and have a tumor, often have 40% of the time, a signature which reveals a new retrovirus . . .[2]

In my estimation, this was a tour de force of scientific investigation. Marry technology to knowledge of biology. This is how science is supposed to be done, an observation raises a question, an investigation is performed, and results are obtained.

The observation was that the genetic defect found in men who were likely to have early-onset prostate cancer was a defect in a gene that created an antiviral defense enzyme.

The investigation involved putting more than a hundred of these tumors on DeRisi's viral detection array and seeing what popped up.

The results showed 40 percent of these tumors contained integrated sequences of a mouse retrovirus.

Science doesn't get much cleaner than that.

DeRisi continued:

Okay, that's pretty wild. What is it? So, we clone the whole virus. First of all, I'll tell you that a little automated prediction told us it was very similar to a mouse virus. But that doesn't tell us too much, so we actually cloned the whole thing. And the viral genome I'm showing you right here? It's a classic gamma retrovirus. But it's totally new. No one's ever seen it before.

Its closest relative is, in fact, from mice. And so, we would call this a xenotropic retrovirus because it's infecting a species other than mice. And this is a little phylogenetic tree to see how it's related to other viruses. We've done it for many patients now, and we can say they're all independent infections.[3]

Silverman, DeRisi, and Gupta did not clone the whole virus in 2006 from a single biopsy. They actually cloned the virus from several different men's biopsies, creating a hybrid, a Frankenstein virus, which never existed in humans.

If they had told the truth in 2006, it would have been much more difficult to discredit our work in 2011. We would never have included the PCR results in our *Science* paper. The protein data, electron micrographs, and serology results from our 2009 Science paper stand as truth to this day. Concealing the truth as they did made the job of the bad guys so much easier.

* * *

A typical retrovirus is eight thousand to ten thousand base pairs.

The VP62 clone was created from tissue samples from three different patients from which they had identified only about three hundred base pairs in the envelope gene covering of the virus, and a couple hundred in the *gag*, or main body of the virus.

How can a synthetic infectious molecular clone be used to characterize a new family of viruses when the sequence of that clone represents less than 10 percent of the natural isolate?

* * *

People often ask me what's the difference between a typical virus and a retrovirus.

The answer is simple.

Many viruses are composed of DNA.

A retrovirus is composed of RNA.

But wait, you say to yourself, aren't pretty much all living things based on DNA?

The answer is, yes, they are.

While nobody knows the answer, the current thinking is that RNA must have been a precursor to DNA, used by the most primitive organisms at the very dawn of life on our planet. Nature, in her unparalleled efficiency, uses nucleic acid building blocks RNA in DNA for essentially all organisms, but those pathogens that are made of RNA, like retroviruses, need an enzyme to transform their RNA into DNA and insert itself into the hosts' DNA blueprint in order for the virus to survive. A retrovirus cannot live or replicate without using the machinery of the host cell.

The enzyme retroviruses used to change their RNA into DNA is called reverse transcriptase. I consider the presence of reverse transcriptase in a disease to be almost smoking-gun evidence of a retrovirus being involved.

If you're a retrovirus, there's just one thing about reverse transcriptase. It's an inefficient enzyme for copying your genetic code.

Reverse transcriptase is prone to copying errors, causing retroviruses to easily mutate into viruses with wildly different genetic profiles. That means there can be large variations between different strains of a single type of retrovirus.

This trait of retroviruses made many scientists believe they could never cause much harm to humans because they seemed to be, well, unstable. Since they are usually easily silenced and crippled, most have dismissed them as noninfectious junk DNA. That may be one reason John Coffin and Harold Varmus both told Frank not to bother looking for disease-causing human retroviruses.

Retroviruses seemed like some biological relic of our far distant past, interesting in the way a platypus intrigues us with its combination of mammal, bird, and reptilian features.

But HIV-AIDS was a wake-up call to the danger of retroviruses. They could be just as dangerous as the typical virus. But HIV really was an outlier among retroviruses. It killed its victims within a few years.

HIV killed relatively quickly for a retrovirus. XMRVs disabled, keeping their hosts alive, but in a state where they could generally not regain their health.

* * *

My life has been about retroviruses since I entered Frank Ruscetti's lab as a technician on June 6, 1983. Frank and I always remember the date, as it was the same date as the D-Day invasion of Europe during World War II.

Teaching viruses how to thrive in cell culture has been my life since that day in 1983. That's thirty-seven years. And you must remember in those days before HIV-AIDS, Frank and Bernie Poiesz were the retrovirology superstars since they'd isolated the very first cancer-causing human retrovirus, Human T-Cell Leukemia Virus (HTLV-1). Bernie was an MD clinical fellow, and Frank was Bernie's mentor. Robert Gallo was the lab chief, getting all the credit and doing essentially none of the work.

Now, HTLV-1 was endemic in Japan, so why did Frank and his team end up first isolating the virus? It's because Frank understood better than any person I've ever known how to use the cell lines to grow viruses and monitor that key enzyme, reverse transcriptase. That's where I learned how to grow viruses, because if you can't grow these viruses, you can't study them.

The best tissue line to grow XMRVs was one called LNCaP (lymph node carcinoma of the prostate), which came from a sixty-two-year-old man whose prostate cancer had spread to his lymph nodes. Why was this such a good cell line? One reason is that it contained defective RNAse-L, the very enzyme that allowed the integration of the sequences detected on the ViroChip.

And it can't really be said I'm somebody who's against infectious molecular clones, since I helped construct the first infectious molecular clone of a HTLV-1 when I worked as a postdoc in the lab of Dave Derse from 1993 to 1994.

Yes, that's right, I've put together infectious molecular clones. I know their strengths and limitations.

Because these viruses like to hide in tissues, they're hard to isolate. They only come out and can be found in the blood when the immune defenses of a person are near zero.

You need cell lines to grow retroviruses, and you need to know what to use.

I don't want to get too technical, but I must explain something because I fear if I don't, you'll just think it's beyond your understanding, and it's not. So, Silverman and Gupta do their investigation, clipping off several genetic sequences of an XMRV, assembling something like three to four hundred base pairs of an estimated eight to ten thousand, and then they seal it up.

Do you understand they've created something new?

Something that never existed in nature?

We used to call these clones Frankenstein viruses, stitched together out of pieces and parts like that terrifying monster created by Dr. Frankenstein.

Then they utilize a technology called "gene-walking," working with enzymes, which do a much better job than reverse transcriptase has ever done in nature, and they make perfect little copies of themselves.

An infectious molecular clone at best is a prototype of an actual virus, akin to taking a baseball and putting it in a copy machine and getting a two-dimensional picture of it. You wouldn't take the copy of a baseball and hand it to the pitcher in the World Series.

Everybody would laugh at you.

But those ignorant of the biology, like critics Bridgette Huber and John Coffin of Tufts University, said it was just fine to use this inferior infectious molecular clone as a stand-in for natural isolates of XMRV. It's like having a scrap from a treasure map with part of the journey, but you're lacking the rest of the map. Where's the starting point? In a treasure map, the important part is where X marks the spot and you can find the hidden riches. In understanding a virus, you need the entire sequence.

Natural isolates of XMRV are very slow growing. The VP62 infectious molecular clone is a nasty little beast, growing at fifteen to twenty times the rate of natural XMRVs, like an invasive weed.

There is one way that XMRVs and VP62 are frighteningly similar.

They could both become aerosolized, floating through the air like seeds from a dandelion on a summer breeze. I have just described the nightmare scenario that keeps virologists awake at night. But there was one other problem that nobody had considered.

If both XMRVs and VP62 could become airborne, we were also infecting our lab workers and scientists. When I began this investigation, I used the antibody test on my blood and those of my collaborators. We were all negative.

Now, I test positive for XMRV, as do many of my collaborators.

* * *

In the wake of the July 22, 2009, meeting, prior to the world knowing about our findings in October of that year, the National Cancer Institute got down to business. They wanted to get the jump on this thing. I commend them for being proactive.

There was just one problem.

At that time, nobody had considered the possibility that XMRV, or its infectious molecular clone, VP62, could become airborne. But the data are all there. Lab workers used as controls were seropositive and contained proteins and immune responses. The big OMG at that meeting was this very realization that XMRVs were more dangerous than HIV, as HIV never showed the ability to spread through the air.

In building 535 of the National Cancer Institute, scientists set up two fermenters in the same room to grow cell lines generated by our team producing the natural isolates of XMRVs and VP62. Now, I've told you that BOTH pathogens can become airborne, but VP62 reproduces much more quickly.

In 2009, we did not know this.

In 2011, however, a team comprised of researchers led by Adi Gazdar from Johns Hopkins University, MD Anderson Cancer Center in Houston,

Texas, and the University of Texas Southwestern Medical Center asked the question of whether human tissue that had utilized mouse biological products in the culturing process might infect other samples, even though they'd never physically touched, in other words, whether the virus could become airborne.

> Six of 23 (26%) mouse DNA free xenograft cultures were strongly positive for MLV and their sequences had greater than 99% homology to known MLV strains. Four of five available supernatant fluids from these viral positive cultures were strongly positive for RT [reverse transcriptase] activity. Three of these supernatant fluids were studied to confirm the infectivity of the released virions for other human culture cells. Of the 78 non-xenograft derived cell lines maintained in the xenograft culture containing facilities, 13 (17%) were positive for MLV, including XMRV, a virus strain first identified in human tissues. By contrast, all 50 cultures maintained in a xenograft culture-free facility were negative for viral sequences.[4]

In plain English, how bad were these findings? Let me break it down for you.

First, if any mouse products were used in the development of a biological product, there was a 26 percent chance it was infected with a mouse leukemia virus.

Second, four out of five possible tests showed a high level of reverse transcriptase activity, meaning the virus was active.

Third, if a tissue sample had never touched a mouse product, but was maintained in a facility that did have such samples, there was a 17 percent chance that this mouse leukemia virus would find its way into the sample. Did it walk from one sample to another on its little viral legs? I don't think so. This was evidence of airborne transmission.

Fourth, if human cell lines were maintained in a facility without any mice-derived products, there was a 0 percent chance they would contain any mouse leukemia virus.

And just in case we'd forgotten some of the basics, the authors were kind enough to provide a little refresher for the casual reader:

> Retroviruses are enveloped viruses possessing an RNA genome and replicate via a DNA intermediate. Retroviruses rely on the enzyme reverse transcriptase to perform reverse transcriptase of its genome from RNA into DNA, which can then be integrated into the host cell's genome with an integrase

enzyme. Retroviral DNA can remain in a latent form in the genome (provi-
rus) or its RNA can be expressed intermittently as infectious virions.[5]

The freak show continues. Retroviruses are made of RNA. In order to replicate in DNA organisms, they need reverse transcriptase. Once transcribed
into DNA, they can go into the genome of that organism and slumber for
extended periods of time or can, under the right circumstances, spew forth
infectious particles.

One would think this series of facts was worthy of a national emergency
rather than trying to destroy our reputations and ignore the suffering of
millions today and in the future.

And my claim that we need to fear animal retroviruses jumping into the
human population? Don't take my word for it. Here's what researchers from
Harvard University told the World Health Organization in a 2012 article
that discussed the risks of using animal tissue for medical purposes:

> Xenotransplantation is any procedure that involves the transplantation,
> implantation, or infusion into a human recipient of live cells, tissues, organs
> from an animal source. This definition may include human bodily fluids,
> cells, tissues, or organs that have had ex vivo [outside the body] contact with
> live non-human animal cells, tissues, fluids, or organs. . . . As with any form
> of transplantation, xenotransplantation carries the potential risk for transmis-
> sion of both known and unknown zoonotic infectious agents of animal origin
> into human recipients and into the wider human population.[6]

This is not a complicated proposition. All it says is when you mix animal
and human tissue, there's a risk that infectious agents present in the animals may cross over into the human cells. The viruses were likely dormant
in the animal, but when put into humans, they may wake up and become
active.

We could also say there's a risk that human pathogens could cross over
into the animal cells. What we're doing is breaking down the barriers nature
has erected so that the pathogens of one species cannot be easily transferred
to humanity and vice versa. We're like children playing with matches and
gasoline, hoping we don't burn down the planet.

And as for my specific fear that retroviruses from animals will cross over
into the human population? Honestly, I don't know why scientists like me
are treated as purveyors of panic when you can go into scientific journals
and read these things for yourself:

> Concern about retroviral transmission in xenotransplantation relates to the potential for "silent" transmission, that is, unapparent infection that may cause altered gene regulation, oncogenesis, or recombination. No exogenous viruses, equivalent to HTLV or HIV, have been found in pigs. However, endogenous retroviruses (part of the germline DNA) have been demonstrated in all mammalian species to date. Endogenous retroviruses that are infectious for human cells in vitro have been detected in many species, including baboons (BaEV), cats (RD114), mice (murine ERV), and pigs (PERV).[7]

Does that last paragraph cause you to lose sleep? It should. In effect it says that retroviruses can affect you in ways that are different from those normally associated with your typical virus.

Transmission is "silent," meaning your body's immune system is not alerted to fight this invader. The pathogen could be said to be "stealth" like a B-1 bomber with radar cloaking and deflecting materials and angles. As if that's not terrifying enough, once established in your body, our little retrovirus will get busy changing your gene expression in unpredictable ways, promoting the development of cancer (oncogenesis) or recombining with other pathogens in your body to create new monsters.

There you find yourself at the doctor's office one day with some funky disease, and your physician is scratching his head as he looks over your results, saying, "Well, you've got something that is kind of like X and kind of like Y. Let's just call it idiopathic X and Y disease." After that doctor visit, you go to the dictionary because you don't know what "idiopathic" is and you find it means "of unknown origin" and you think to yourself, "Maybe my doctor's the idiot."

How can these people not know?

Where's the slightest bit of scientific curiosity?

* * *

Let's talk about Silverman's VP62 molecular infectious clone. When Frank and I worked with it, we estimated the VP62 infectious molecular clone grew fifteen to twenty times more quickly than natural isolates of XMRVs. Frank was terrified by this finding, and little terrifies him. Even Ebola didn't concern Frank as much as Silverman's VP62 clone.

Silverman and Gupta created something that had never previously existed in nature. Is it any surprise that when they told labs like Abbott to create a test based on their VP62 plasmid, they weren't able to find anything like it in their patients?

And so, if you have one fermenter in a room at the National Cancer Institute with XMRV and another with the VP62, what do you think is going to happen? Who wins that evolutionary struggle? We know that XMRVs can become airborne because of the Adi Gazdar paper I cited earlier. Wouldn't the man-made XMRV called VP62 do the same thing?

Let's say we have transmission going both ways. The fermenter with XMRV contaminates the VP62 and vice versa. But since VP62 grows so much more quickly, it will easily wipe out any traces of natural XMRVs or, worse yet, recombine to create new Frankenstein XMRVs.

* * *

We never had VP62 in our WPI labs, which is why we were able to get accurate results, free of any VP62 contamination. Our cultures were contaminated in the NCI labs, and we proved it by single cell cloning patient samples, which I showed in that September 22, 2011, debate.

I was always trained to keep the samples separate, in case of just such an occurrence.

I have a good deal of sympathy for those who were upset that Silverman's VP62 synthetic infectious molecular clone contaminated their lab, like it contaminated those in building 535 of the National Cancer Institute.

In Silverman's defense, he did not know that the VP62 synthetic infectious molecular clone could become airborne when he created it in 2006. We only learned that in 2011, five years later.

However, the VP62 plasmid contamination issue has nothing to do with the question of whether XMRVs are infecting the human population.

* * *

"Joy really feels bad about all this," Silverman said to Frank Rusectti at a break during the 15th International Conference on Human Retrovirology, HTLV, and Related Viruses, held in Leuven, Belgium, June 4–8, 2011. "Joy" referred to J. Das Gupta, Silverman's research partner.

What Joy felt bad about was the fact that they'd taken sequences from three different patients to create their infectious molecular clone, which clearly meant they did not have an accurate sequence of our XMRVs that they contaminated with their VP62. Yet they reported their sequences as "highly similar" to natural isolates of XMRV in our *Science* paper. They felt bad about it? If they felt so bad, why did they wait to tell us until the summer of 2011? To this day they let me take the fall, knowing full well that countless lab workers, including me, were being infected and are now developing devastating neuroimmune disorders and cancers. I am infected, and many of my former colleagues are, as well. All because the government was not interested in retrofitting laboratories to Biosafety Level-3 precautions, the same safety standard used for HIV. Animal caretakers and student lab technicians are also being injured from this negligence.

These cowards simply ignored the multiple lines of evidence we had from the beginning showing our data was not the result of VP62 contamination.

Foremost among that evidence was Frank's presentation titled "Development of XMRV Producing B Cell Lines from Lymphomas from Patients with Chronic Fatigue Syndrome." We had theorized that one of the long-term effects of viral infection would be an increased rate of cancer, specifically B-cell lymphoma, otherwise known as Non-Hodgkin's lymphoma. Among the general population, the rate of non-Hodgkin's lymphoma was 0.02 percent, but among sufferers of ME/CFS, the rate was close to 5 percent. Frank had written in his presentation:

> Additionally, development of cancer coincides with an outgrowth of gamma delta T cells with specific clonal T-cell receptor gamma. We hypothesized that infection with XMRV and/or other viruses can trigger a dysregulated immune response, which favors the development of B-cell lymphoma.[8]

It was one thing to claim that VP62 had contaminated various research labs; it was quite another to claim it had found its way into long-time sufferers of ME/CFS. Harvey Alter's work, especially, disproved this idea, as his samples had been collected in the mid-1980s, prior to the creation of VP62 in 2006. Similarly, John Coffin's idea that the recombination to create XMRV had happened in a lab at Case Western University in the 1990s also did not make any sense. If simply culturing viruses could generate infectious recombinants in two weeks (reverse transcriptase was actually detected after ten days), how could Coffin proclaim in 2013 that XMRV was a freak of nature, a once-in-creation occurrence so fantastical we might as well call it the "immaculate recombination." If it happened once in a culture, it could easily happen again.

Silverman wanted to be done with the issue and was against letting Frank and me do the necessary testing to determine whether VP62 had somehow found its way into our samples. On July 7, 2011, Frank sent a scathing email to Silverman:

> Dear Bob:
> I find your answer disingenuous to say the least. *The source of contamination is of interest to us, but having that answer will not change the fact that Fig. 1 is erroneous.* (italics added) Depending on the source of contamination, it will determine whether all the figures or just fig. 1 is discredited. All statements in this field about contamination pinpoint the problem, science demands that we do so also. *We may never figure out the source of the contamination, but we need to make public aware that the interpretation of figure 1 is just plain wrong.* (italics added) The reality is since the publication of the paper in 2009 nobody has believed fig 1 is correct so why the rush to publish before doing the experiments we suggest. Second, we all suspect (you may already know thus the comment you said to me in Belgium that Joy feels awful about it) the source of contamination is your lab . . .[9]

Is all of this becoming clear?

Silverman had put together a synthetic, this VP62 molecular clone, which was fine as a prototype to possibly help us understand things about the virus. The problem was he didn't tell us he took pieces from three different viruses and it wasn't even close to a fully sequenced eight to ten thousand base pairs of the virus.

Using that crappy clone, they developed all these crappy tests.

As the programmers like to say, "Garbage in, garbage out."

When we did our initial tests, we saw that the PCR was overlapping in two key virus regions, but the bands were different sizes. If the bands were different sizes, it meant there were different nucleic acids in our strains. Yes, we had something close to Silverman's virus, a different strain, or so we thought at the time.

Let me tell you a difference between Frank Ruscetti, Judy Mikovits, and Robert Silverman.

Frank and I have the courage and integrity to publish all the data, even the part that might not make sense at the time. Not only did Robert Silverman not have the integrity or courage to admit the mistakes made in his laboratory, but he has done nothing for nearly a decade as my career was ruined and millions have continued to suffer and die.

Figure 1 of our original *Science* paper was a picture of the PCR positives in our study based on Silverman's VP62 clone. These were the patient samples that were supposedly PCR positive for the virus we isolated. It was the basis of the title saying it was XMRV based on Silverman's strain. Figure 3 supports that with protein proof detected with antibodies, e.g., 1118, clearly negative by PCR and protein positive in figure 3.

Why would the PCR variations matter when we know that HIV has more than a hundred strains? We had the antibody and proteins that showed these isolates were the true XMRVs.

Was there really anything else that mattered?

I think what really mattered is that our guardians of public health simply didn't want to admit the terrible problem they'd created.

And the unique characteristics of VP62, its ability to become airborne and quickly overgrow the XMRV virus from which it had been created, made it the perfect assassin for my career and the lives of millions who would continue to search for answers.

CHAPTER EIGHT

My Identity Stolen in Vaccine Court

In George Orwell's classic dystopian novel *1984,* there's a process by which an enemy of the state is expunged from memory. It's called becoming an "unperson," and Big Brother will simply wipe away all traces of your existence. The population has learned not to question the sudden disappearance of people, or radical changes of government policy, such as abrupt changes of alliance in an endless war. One day the bitter, longtime enemy will be East-Asia and the next day it will be Eurasia, and the government authorities will pronounce East-Asia has always been the longtime ally. There must forever be an enemy to hate, and the vague fear that if one doesn't behave according to written and unwritten government dictates, you'll wake to find you've been erased from memory.

At the start of my journey in 2009, I didn't realize I was an unconscious follower of the government and scientific propaganda regarding vaccines. In retrospect, I should have questioned more. But I find that most of those who now find themselves on the so-called "anti-vax" side started off in much the same way.

I was simply following well-established clinical practice when I mentioned on a Nevada television show in 2009 that if a retrovirus was involved in ME/CFS and autism, we should treat it with the same precautions we did an HIV infection. XMRV might not be a killer like HIV, but it certainly did a lot of damage. Any immune stimulation could cause the virus to replicate out of control, causing it to go from a virus that the immune system had silenced to one that was rampaging through the body. Children born to HIV-positive mothers immediately go on antiretroviral therapy prior to any immunizations.

I did not expect the vicious response that followed that TV show.

Maybe you think I should have taken a hint and stopped talking about it.

Did Columbus stop talking about the Earth being round when he met opposition?

Did Galileo stop talking about the Earth revolving around the sun when the clergy told him he was contradicting the Bible?

Scientists aren't worried about their popularity. They worry if their ideas are true.

When nobody was able to give us a good explanation as to why we shouldn't follow similar precautions with XMRV as we did with HIV, we followed the most stringent of safety precautions and kept investigating.

A scientist knows that a good question leads to other questions.

As I've said before, that question led me to more fully understand the threat posed by using animal tissues in medical products. I also began to understand the simple question I'd asked could also be asked of many of the products used in vaccines and other medical products.

If science had missed this problem, as we'd missed the problem with lead in gasoline and other products for many years, I wondered why the legal system had not taken on this challenge.

I was about to get an education as to how the pharmaceutical industry had intervened in our laws to make sure a strong case would never be made against vaccines.

Vaccines would be touted as such a shining example of scientific progress that hard questions about them could never be asked in a regular courtroom.

* * *

As a scientist, I'm a big fan of data and evidence.

If you pose a hypothesis, a scientist proves or disproves it with data. Honestly, that's how some of the greatest discoveries are made. Two statements of alleged fact are made, but they can't be reconciled. It's the apparent mistakes, or the two things that can't both be true, that draw our attention.

Here are two things that cannot both be true:

Vaccines are as safe as sugar water.

Vaccines carry such risk that nobody is liable for injury.

Claims may only be brought in a special court, where the parents of vaccine-injured children face off against Department of Justice attorneys. This

Getting my doctorate in bio-chemistry and molecular biology from George Washington University in 1991. This is during the "hooding" ceremony and Frank, as my sponsor, is directly behind me in his doctoral robes.

Frank with Dan Stevens, a good friend and local radio disc jockey with a morning show. He'd usually open his program by saying, "It's five in the morning and I know Frank and Judy are probably in the lab trying to cure cancer."

Me with the championship softball team I coached at Fort Detrick in 1990.

Building 567 was where we had our lab. The Bio-Safety Level 3 lab was on the corner of the third floor. Frank and I would usually race each other up those stairs every morning to be first into the lab. The large ball structure known as the "eight-ball" partially visible behind the building is where they gassed animals with chemical and biological agents. Behind the eight-ball was the building where they tested anthrax.

With some of my colleagues when I served as Director of the Lab of Anti-Viral Drug Mechanisms at the National Cancer Institute from 1999 to 2001. In announcing my appointment, the Director said I was "an accomplished virologist with a number of important publications," and he expected me to continue their "tradition of excellence." I think I did.

An aerial view of the Animal Farm division at Fort Detrick.

The eight-ball where they once gassed animals with chemical and biological agents is now listed as a historic landmark and cannot be torn down.

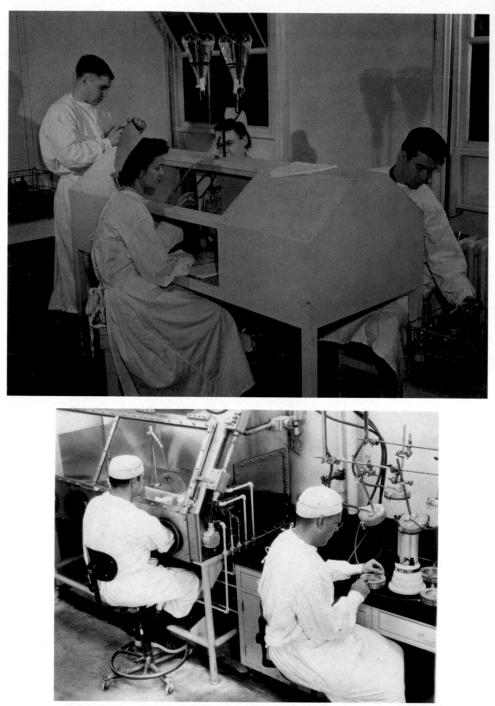

Much of public health research has been conducted under military control. Men and women both performed this patriotic work, but I wonder if we truly understood the forces we have unleashed through the use of animal tissue.

Kent was happy that since his name went first, we'd be shelved right next to famed scientist Stephen Hawking in the Science section. What a nerd!

My husband David and me.

The cartoonist Ben Garrison was outraged by Australia's treatment of Kent and drew this cartoon in 2018 to highlight the rise of "cancel culture" on many important topics.

Cartoonist Ben Garrison came back with another cartoon when he heard about our new book. I respect people who don't get scared by bullies.

When Kent was banned from speaking in Australia he threatened to sneak into the country in a disguise and using a phony name. Luckily, his friends and family were able to convince him to abandon the plan.

How did mouse retroviruses get into humans?

"One of the most widely distributed biological products that frequently involved pieces of mouse tissue, at least up until recent years, are vaccines, especially vaccines against viruses . . . It is possible that XMRV particles were present in virus stocks cultured in mice or mouse cells for vaccine production, and that the virus was transferred to the human population by vaccination."

(Frontiers in Microbiology, January 2011)

I find the best explanations are usually the simplest. We should not be injecting animal tissue or aborted human fetal tissue into people. It should stop **NOW**!

special "Vaccine Court" was established by law under the 1986 National Childhood Vaccine Injury Act.

Many of you are probably learning about this court for the first time. The Act removes liability from pharmaceutical companies for any injuries or deaths caused by their childhood vaccines. It also only allows discovery of pharmaceutical company documents with the approval of the court. This is dramatically different from the typical products liability action in which the company must turn over all relevant documents. In addition, each case is unique, meaning that if a vaccine has been found to cause a certain injury in one case, that cannot be relied upon in any way in a subsequent case.

In addition, parents may not seek documents on the safety of vaccines from the companies that produced them, unless given specific permission from the court. And if one parent is successful in making a claim of harm from a certain vaccine, that information is then hidden from others who want to make a similar claim.

If vaccines were as safe as sugar water, then there would be no need for such a system.

Is there a soda court, where claims against soft drinks are made, defended by Department of Justice attorneys?

* * *

Kent and I have done extensive research into this so-called "Vaccine Court," and I'll briefly recap what we've found.

The truth is that vaccines are so dangerous that the liability from one shot, that earlier version of the DPT shot for which my attorney Mike Hugo worked so magnificently, was so great that the pharmaceutical companies didn't want to face such scenarios in the future.

One of the experts Kent interviewed was Stanford Law professor Nora Freeman Engstrom. Professor Engstrom is one of the few academics who has examined the chain of events that led to the formation of the court in 1986 as well as how the court has functioned in the ensuing decades.

> Engstrom reviewed the case of Anita Reyes, a young girl living near the Mexico border in the 1970s who contracted polio as a result of a dose of a Wyeth Laboratories polio vaccine. The case that resulted, *Reyes v. Wyeth*, heard in the Fifth Circuit, suggested that between victims and vaccine manufacturers, the manufacturers should bear the loss.[1]

The loss in the Reyes case was followed by a series of calamities for the vaccine makers. Forty-five million Americans took a defective Swine Flu vaccine in 1976, with several hundred developing Guillain-Barré syndrome and suing for their injuries. The DPT vaccine that Mike Hugo had been attacking became a major source of liability for the vaccine makers. As Stanford Law professor Engstrom detailed in a later article:

> As the number of lawsuits ticked upwards, so did the manufacturers' dismay. In 1984, for example, Lederle's President went on record declaring that "[t]he present dollar demand of DTP lawsuits against Lederle is 200 times greater than our total sales of DTP vaccine in 1983." . . . Another vaccine manufacturer—Connaught Laboratories—faced a similar plight, as suits filed against it in 1985 and 1986 sought a combined billion dollars in damages.[2]

The two main justifications for creating a "Vaccine Court" were to speed the process along so that the parents of a vaccine-injured family could quickly receive their money and to remove the antagonism so often present in personal injury cases against a large corporation.

The US Government Accountability Office (GAO) looked at the program and found that both parents and those who administer the program had come to similar conclusions about the problems with the Vaccine Court:

> While [the Program] was expected to provide compensation for vaccine-related injuries quickly and easily, these expectations have "often not been met." A leader in the parents' lobby, instrumental in the Act's passage, has concluded that the VICP's administration has constituted a "betrayal of the promise that was made to parents about how the compensation program would be implemented." And the man who served for over two decades as the VICP's chief special master, has publicly lamented: [L]itigating causation cases has proven the antithesis of Congress' desire for the program."[3]

Kent and I try to cover our bases. Not only did he interview one of the few legal academics who has looked at the workings of the Vaccine Court, but he also interviewed the former chief special master of the program for more than two decades, Gary Golkiewicz.

The chief special master of the Vaccine Court is equivalent to the chief justice of the United States Supreme Court.

As of this writing, the Vaccine Court has paid out more than four billion dollars in claims to children who have been injured by vaccines.

* * *

Kent found the former chief special master, Gary Golkiewicz, surprisingly open to sharing his experiences. When asked about how the program could have been better, Golkiewicz had strong opinions.

"We're operating today with an Act designed to handle the DTP shortage, which we no longer even give. The original Act had I believe six vaccines, but now I think we're up to eighteen or nineteen. So that original language is being applied to the HPV [Human Papilloma Virus] vaccine and also the flu vaccine, which is the number one source for work in the program. And it's not effective. And that's what's causing all the frustration from the parents' side, which is absolutely correct. The argument that the program takes too long, it's too litigious, and it's not quick justice is correct. There's a tie-up in the courts right now. They expected a hundred and fifty cases a year and last year they had over nine hundred."[4]

Kent brought up the challenging question of whether the alleged concealment of vital information from the Vaccine Court, such as the Simpsonwood Conference of 2000 and cover-up of Generation Zero data on mercury exposure, the allegations of destruction of data in 2004 by CDC scientist, Dr. William Thompson regarding the MMR (measles-mumps-rubella) vaccine and autism, and the 2009 embezzlement charges against Dr. Paul Thorsen, one of the main authors on several studies showing no link between vaccines and autism. To his credit, Golkewicz did not duck the question.

"I would say this in answer to your question," said the former chief special master. "The special master, like any judge, relies upon information. If that information is not correct, it would obviously impact the information the judge is considering and could potentially impact his decision. In that respect, the special masters are no different than any other judge. Your decision is only as good as the information you get."[5]

The question that is raised, but not answered, is who controls the information given to the Vaccine Court.

I was going to get an education on how these gatekeepers of information acted when qualified scientists stepped up as expert witnesses addressing vaccine safety in the Vaccine Court.

I would not be impressed with what I learned. In fact, I am outraged.

* * *

It's probably important we spend a brief amount of time explaining the Simpsonwood Conference of June 7 and 8, 2000. Robert F. Kennedy, Jr. did a masterful job explaining what happened in a June 16, 2005, article in *Rolling Stone* and *Salon* titled "Deadly Immunity." My coauthor spent an entire chapter on the conference in his fine book, *INOCULATED: How Science Lost its Soul in Autism.*

Fifty-three individuals met at the Simpsonwood Conference Center over the weekend of June 7 and 8, 2000, to discuss troubling findings regarding mercury and aluminum in the new vaccine schedule and the link that kept coming up in the data that the more vaccines you gave a child, the more likely they were to have a neurological problem.

In science, this is known as a "clue," which is likely to help you better understand an issue.

The meeting was chaired by Dr. Walter Orenstein, who at the time was director of the National Immunization Program at the Centers for Disease Control (CDC). Of the fifty-three individuals present, fifteen were members of the CDC. There were fourteen individuals with academic appointments, but who also often had overlapping advisory roles with certain public health agencies. Twelve individuals directly represented the National Immunization Program. Five were representatives of the vaccine companies. Four were representatives of the National Academy of Pediatrics. Three individuals were not medical doctors: Paula Ray, the project manager for the Northern California Kaiser Vaccine Study Center; Ned Lewis, the data manager for the Kaiser group; and Wendy Heaps, the health communication specialist for the National Immunization Program. Two individuals represented state health agencies, and two represented foreign entities, specifically the European Agency for the Evaluation of Medicinal Products and probably the most eminent member of the group, Dr. John Clements, who represented the World Health Organization (WHO).

In looking at the data they'd assembled, both for what they were able to conclude about mercury exposure and the related question of what they didn't understand about the use of aluminum salts, lead study author Thomas Verstraeten summarized the findings:

> The bottom line to me is you can look at this data and turn it around and
> look at this, and add this stratum, I can come up with risks very high. I can
> come up with risks very low, depending on how you turn everything around.
> You make it go away for some and then it comes back for others. If you make

it go away here, it will pop up again there. So, the bottom signal is, okay, our signal will not go away . . .

Personally, I have three hypotheses. My first hypothesis is it is parental bias. The children that are more likely to be vaccinated are more likely to be picked up and diagnosed. Second hypothesis, I don't know. There is a bias I have not yet recognized, and nobody has yet told me about it. Third hypothesis, it's true, it's thimerosal. [mercury][6]

Let's put the best face possible on this assessment. They were noticing an effect and didn't know if it was real. I think you could make the argument that if the information wasn't clear, they didn't want to panic the public over something that might end up being false.

However, at the very least, if you had the slightest concern for the safety of the public, you'd want to know if you were harming children, wouldn't you? But that doesn't seem to have been the consensus of public health officials and members of the pharmaceutical industries. The dominant view was expressed by Dr. John Clements, who, given his position at the World Health Organization, was likely to command the most respect. In his summation, Clements said:

It has been a great privilege to listen to the debate and to hear everybody work through with enormous detail. And I want to congratulate, as others have done, the work that has been done by the team. Then comes the BUT. I am really concerned that we have taken off like a boat going down one arm of the mangrove swamp at high speed, when in fact there was no discussion early on about which way the boat should go at all. And I really want to risk offending everybody in this room by saying that perhaps this study should not have been done at all.

Because the outcome of it could have, to some extent, been predicted. And we have all reached this point now where we are leg hanging, even though I hear the majority of the consultants say to the board that they are not convinced there is a causality direct link between thimerosal and various neurological outcomes. I know how we handle it from here is extremely problematic.[7]

Are you reading his summation the same way I am? That maybe this is a study that should not have been done? In my entire life I've never known a scientist to argue against obtaining knowledge.

And it isn't the harm to children he's worried about, but how this information can be managed. Honestly, it seems they've done a pretty good job of managing the information in the nearly two decades since that meeting was held.

My coauthor, Kent, isn't surprised at what's happened to me over the years. "They knew somebody like you would eventually come along, Judy," he says. "Somebody who wouldn't be bullied, no matter what they did, and wouldn't give up."

Maybe that's true, but it makes me sad.

I'd grown up respecting the honesty and integrity of scientists involved in public health.

* * *

I was asked to work as an expert witness in the vaccine court by several longtime vaccine court attorneys. In 2015, Frank and I formed a consulting company, Mikovits and Ruscetti Consulting, Inc., or M.A.R.C. Inc. In one of our interactions with the court, Frank and I were asked to describe our background. We wrote:

> Drs. Mikovits and Ruscetti changed the practice of immunology and medicine arguably more than any two individuals since 1980. As pioneers in translational medicine, we have worked our entire careers together in teams with MDs, nurses, nurse practitioners, Ph.Ds. Our passion throughout our careers is to work together to translate discoveries as rapidly as possible and free from bias and conflicts of interest which is why we succeeded. It is *because of* this approach and our discoveries as part of these teams, we changed the paradigm of treatments not only for HIV/AIDS but also for cancer, autoimmune disease and neuroimmune disease (Autism, ME/CFS, MS. ALS etc.) Our discoveries and treatments based on them and developed by our teams are credited with saving millions of lives. We have highlighted some examples as they address questions 10–14 regarding the source of our knowledge of vaccine manufacturing, its contaminants and the immune related adverse events associated with vaccine components, excipients and contaminants.[8]

These are some of the accomplishments we listed for Frank, who for thirty-nine years worked at the very center of government science investigating human health:

- In 1978, Dr. Ruscetti's team discovered interleukin 2, then called T-Cell Growth factor.

- Dr. Ruscetti's team received an award from the society for immunotherapy in cancer for interleukin 2. It was a team award for MDs and PhDs who collaborated on the first clinical trials.
- In 1980, Dr. Ruscetti's team isolated the first pathogenic human retrovirus HTLV-1, which is the causative agent of adult T cell leukemia and associated with autoimmunity, acquired immune deficiencies, and cancer.
- Dr. Ruscetti's teams also discovered additional pathways of immune pathogenesis and the molecules that regulate these pathways including IL-5 and IL-15. These discoveries were key to the understanding of T-cell immunology, cytokine regulation of inflammation, and efficacy and toxicity of immune therapy.
- Dr. Ruscetti's teams paper on T cell growth was awarded the American Association of Immunologists second most important publication of the first hundred years of publication.
- In that same time frame, Dr. Ruscetti developed a cell line from a patient with acute promyelocytic leukemia (APL). This allowed the demonstration that several biological agents could induce complete differentiation of APL. Retinoic acid induced differentiation became the standard of care.
- Dr. Ruscetti's teams first associated HTLV-1 with Adult T cell leukemia (ATL) in 1980. For this, we received an award for the "Development in the field of Human Retrovirology" from the International Retrovirology Association and recognized as coauthor of one of the top thirty papers published in the hundred years of the National Academy of Sciences.
- In 1991 Dr. Ruscetti was awarded the NIH [National Institutes of Health] Distinguished Service Award "in recognition of fundamental co-discoveries of interleukin-2, the first human leukemia virus, and for the discovery of hematopoietic regulatory activities of transforming growth factor beta."

As for me, I had a similar list of accomplishments.

- Dr. Mikovits was part of the team in 1980 who purified interferon alpha. It was this interferon alpha which was the first immune therapy for cancer, a curative therapy for hairy cell leukemia.
- In 1986 Mikovits took a job at Upjohn Pharmaceuticals in Kalamazoo, Michigan. The goal of her project was to provide

experimentation demonstrating that the GMO bovine growth hormone did not harm human cells. (It did.) A second project with Dr. Ruscetti involved proving if Upjohn's biological drug known as ATGAM (made with human blood) was contaminated with HIV. The manufacturing process did inactivate HIV, thus removing the risk of HIV transmission.

- Mikovits was awarded graduate student of the year in 1991. This work changed the paradigm of HIV/AIDS treatment and has saved millions. It also showed that HIV does NOT CAUSE AIDS because a genetically resistant person can be infected with HIV and never develop AIDS. It was the immune response to HIV, not the infection, that caused the damage.

- In 1992 Mikovits and Ruscetti began a collaboration with Stephen B. Baylin, MD, of Johns Hopkins University to understand how retroviruses dysregulate the DNA methylation machinery. This seminal 1998 publication directly impacted therapeutic protocols for AIDS and cancer, showing how retroviruses dysregulate cellular gene expression. Publication took four years because it went against all known dogma. A key aspect of this work was the discovery that infectious virus was NOT required to dysregulate DNA methylation.

- Between 1994 and 1998 we published data showing that the inflammatory pathway NFKB was dysregulated by HIV infection. Again, this discovery made no sense given the dogma of the time concerning these two pathways.

- In 1999, Dr. Mikovits was appointed director of the Lab of Anti-Viral Drug Mechanisms in the Developmental Therapeutics program of the National Cancer Institute. The goal of the lab was to develop therapeutics for AIDS-associated malignancies.

- Dr. Mikovits and her team worked in collaboration with Lou Staudt, MD, PhD, who pioneered the development of gene expression profiling technology to define molecular subtypes of lymphoma for targeted therapies.

- In May 2001, Dr. Mikovits left the National Cancer Institute to work at EpiGenX in Santa Barbara, CA, and direct their Cancer Biology program and was named chief scientific officer in 2005. This program developed and implemented high throughput screening technologies for the diagnosis and treatment of cancers caused by the dysregulation of DNA methylation machinery by environmental toxins and retroviruses. The company was acquired in 2006 by a large pharmaceutical company that still uses the drugs Mikovits developed.

- In 2006, Dr. Mikovits designed and developed the research program of the first neuroimmune institute in the world at the University of Nevada (Reno), using a systems biology translational/team approach.
- In 2009, Drs. Mikovits and Ruscetti were part of a team in collaboration with the Cleveland Clinic who isolated a new family of retroviruses then called XMRV and associated then with chronic fatigue syndrome/myalgic encephalomyelitis (CFS/ME). This work was published in the journal *Science*.

Certainly, our combination of nearly sixty years' experience in government scientific research should be enough to qualify us to offer a medical opinion on whether an individual has suffered a vaccine injury. In answer to a question posed specifically about vaccines and what might be some misconceptions among the general public, we explained our opinion that:

> Vaccines are immunotherapy and carry the same risks of immune-related adverse effects as other biological drugs and immunotherapies. Our work changed treatment paradigms of acquired immune deficiencies and cancer by defining at the molecular level genetic and epigenetically susceptible populations (that is, those most likely to suffer Immune Related Adverse Events (IRAE). The lay person doesn't appreciate that vaccines are not single-ingredient pure (antigen only) drugs, but rather multi-ingredient preparations specifically designed to intensively challenge the immune system in the manner of an actual disease. Contaminants from the manufacturing process (called excipients, CDC Excipient table 2 EXH5) that are not removed by the manufacturing process and remain in the vaccine in significant amounts.[9]

My coauthor, Kent, is always telling me I need to simplify things and break it down so that the lay reader may more clearly understand my points. But realize that to me, the above paragraph is perfectly clear.

Vaccines work by stimulating the immune system. We understand that stimulating the immune system is a way in which we might treat cancer. However, when we investigate immunotherapies for cancer, we're carefully monitoring the response of the immune system to our interventions.

However, with vaccines we've been stimulating the immune systems of individuals for the better part of a century with virtually NO monitoring of how their immune system is responding.

Each intervention increases the likelihood of an adverse event.

Increase the number of interventions/vaccines, and you dramatically raise the likelihood of adverse reactions. That's simple logic. Add to that complexity the fact that vaccines contain multiple ingredients, and you realize what a shaky foundation upon which we've built our current medical system.

Further in the document Frank and I noted:

> MMR, polio, and varicella are live attenuated vaccines. The contaminants and excipients include human MRC5 cells, Human WI-38 lung cells, monkey kidney cells, guinea pig cell cultures and bovine serum. Live viral vaccines are all grown in human and animal cells lines and these animal and human cell lines contain human and animal retroviruses (adventitious agents which can recombine to generate new infectious retroviruses during the manufacture.) In addition to the animal and human retroviral contaminants, the carcinogen formaldehyde, antibiotics which dysregulate the GI [gastro-intestinal] and nasopharyngal microbiomes, glutamate, and bio-incompatible contaminants including nickel and chromium (EXH 6) can synergize in toxicity and the development of neuroinflammatory, neurodegenerative and neuroimmune diseases and cancer which can become clinically apparent decades later.[10]

Let's talk about what's become clear to me about vaccines. The vaccines contain human cells, specifically MRC5 and WI-38 cells, from aborted babies.

My friend, the late Dr. Jeff Bradstreet, was extremely concerned that injecting human DNA into the bloodstream of children would eventually integrate into the DNA of those children and cause damage. Dr. Theresa Deisher, who obtained her PhD in molecular and cellular physiology from Stanford University and has twenty-three patents to her name, also has similar concerns and has testified about them.

For many it is a moral issue to use aborted human fetal tissue, but also a scientific problem. The fact is we do not know the long-term consequences of injecting human DNA into the bloodstream of young children (or adults, for that matter). If the human DNA also contains or activates latent retroviruses, these can also generate reverse transcriptase, which will allow for genetic rearrangement, which we know promotes the development of cancer.

Here is a section of testimony given by Dr. Deisher at Georgetown University in September of 2008:

> How might the human DNA contaminated vaccines contribute to human disease? First, there is the potential for the contaminating DNA to be mixed with our own genes by a process called homologous recombination. Homologous recombination is an established biologic phenomenon in which a segment of a cell's DNA is substituted by another segment of DNA that is similar. This can occur during cell division or DNA repair.
>
> Homologous recombination occurs naturally to create genetic diversity in our offspring and is also conveniently harnessed by scientists to introduce experimental DNA into cells or animals. We do not yet know if this occurs with the contaminating human DNA found in some of our vaccines, and if so, to what extent. Imagine the potential consequences of human DNA from a vaccine, a vaccine that is given to children at an average age of 15 months, being incorporated into a child's developing brain. One does not need to be a rocket scientist to know that this potential has to be studied.
>
> In addition to the potential for homologous recombination, DNA is known to be a powerful immune stimulant. Diseases like graft versus host, juvenile (type I) diabetes, multiple sclerosis, lupus and some forms of arthritis are what are called auto-immune diseases. These are diseases driven by immune attack from our own immune system on our own organs, a system normally responsible to attack invading bacteria and pathogens. Targeted self-destruction, if you will.
>
> Science does not yet know, except for graft versus host disease, what triggers the auto-immune attack. We certainly lack studies that have examined the relationship between immune responses to human DNA containing vaccines and auto-immune diseases.[11]

We'll leave aside for the moment the moral issue of using aborted human fetal tissue. There are likely to be significant biological problems from the use of this tissue.

When we move to the animals, the problem becomes much worse. We know with scientific certainty that one of the most destructive factors in human history is the jumping of animal viruses into humans. The Pulitzer Prize-winning author Jared Diamond in his book *Guns, Germs, and Steel* makes the argument that the reason European contact with Native Americans was so devastating is that we had become infected with viruses from our domesticated animals. These came from our horses, our cows,

our goats, our dogs, and maybe the rats that bred in our urban areas, especially combined with the fact we lived in large, concentrated cities with poor
sanitation.

That's a perfect recipe for disease creation.

Therefore, we had a group of disease-carrying Europeans arriving on
American shores with superior weaponry against a population with none of
these advantages.

I make the case that what we have done in the laboratory mimics the
worst of what happened in European populations and animals in the Middle
Ages and then traveled across the ocean. The sad possibility is that our children and immune-compromised are the Native Americans in this scenario.

We are mixing animal and human tissue in laboratory cultures, then
injecting them into human beings in a way that bypasses their traditional
defenses, such as stomach acid breaking down pathogens. Antibiotics, which
we give with alarming regularity, are known to dysregulate the bacteria in
our digestive system, and there's strong evidence of harm from many of the
chemicals used in vaccines.

I understand that to many my concerns might be the first time they've
considered them. However, on consideration of what I've proposed, have I
said anything that seems to be unsupported by the scientific facts?

Have I made a single remark that convinces you my opinion should not
be heard in a courtroom and the truth or falsity of it considered?

* * *

In the cases Frank and I worked on in vaccine court we reviewed injuries
other than autism. One of them was the progression of a young child's juvenile diabetes.

In a letter to the attorney in charge this case, Frank wrote:

> I previously submitted a report in this case, which supported your petition
> claim that DTaP vaccination administered on February 14, 2013 "**signifi-
> cantly aggravated J.B.'s underlying genetic autoimmune susceptibil-
> ity causing the development of T1DM [type 1 diabetes mellitus] that
> but for the vaccine would probably never have happened.**" My original
> report highlighted my 40 years of translational research expertise beginning
> with the discovery of several cytokines and their signaling pathways key to
> the development of the adaptive immune response and the mechanisms of
> pathogenesis of dysregulation of the balance between the innate and adaptive

immune response to infection in the development of chronic inflammatory, autoimmune, and cancer.

As I wrote in my original report, "**we have learned the immune system is not static, it changes with age according to environmental toxins, infections, and vaccinations. Vaccines are immunotherapy. That is, they are designed to alter the immune response to an antigen/infection. We have long recognized that the reasons some individuals react badly to immunotherapies, including vaccines, while the majority of people who are treated/vaccinated do not, has much to do with the recipient's genetics and the status of the recipient's health and immune system at the time of administration of the immunotherapy/vaccination.**"[12]

In another case Frank and I reviewed the case of a thirteen-year-old young woman who appeared to suffer from postural orthostatic tachycardia syndrome (POTS) caused by an HPV (human papilloma virus) vaccine and a hepatitis A shot. (Since she is a minor, she is simply referred to by her initials, CM.) POTS is a condition in which otherwise healthy individuals find that upon standing they suffer from dizziness, lightheadedness, fainting, and tunnel vision. Testing generally shows an abnormal heart rate upon standing, showing a clear biological basis for the condition. A 2012 Mayo Clinic review listed vaccines as a possible cause of the condition.

In our report we wrote:

> CM was a healthy 13-year-old training for a triathlon at the time she was vaccinated with HPV and Hepatitis A vaccine simultaneously on March 20, 2013. Five weeks later CM experienced severe migraine headaches and a syncope event which was distinct from the two previous events in its duration and severity. Quoting Rule 4 Document 15 at 7 "On January 9, 2014 CM reported to Dr. [name removed] for abdominal pain. Dr. [name removed] noted that prior to May 2013, CM had minor syncopal episodes which became significant in May, during which time she had complete loss of consciousness for 20–30 minutes and became bradycardiac and hypoxic with abnormal eye movements."
>
> At that time five weeks after the administration of the HPV and Hepatitis A vaccines CM was diagnosed with two autoimmune/inflammatory diseases: POTS and atopic dermatitis. POTS and atopic dermatitis are interrelated in that they are both immune mediated. At the same time, she is diagnosed with sinusitis . . .
>
> In summary, while CM had risk factors as evidenced by a family history of drug sensitivities and allergies, CM was never symptomatic or diagnosed

with POTS or AD prior to 5 weeks after the March 20, 2013 vaccination. The timing is well within the timeframe for the development of HPV vaccination syndrome (ASIA/HANS). Thus, it is our opinion to a reasonable degree of medical certainty that HPV vaccination caused CM's POTS, atopic dermatitis, fatigue, joint pain, lack of sleep, severe headaches, lack of ability to concentrate, fogginess, lack of sharp memory, severe anxiety, stress, and panic attacks.[13]

This case absolutely broke my heart. Imagine being a thirteen-year-old girl training for a triathlon. Is this a future Olympic medalist? Seriously, how many thirteen-year-olds do you know that are training for a triathlon?

You are incredibly fit but then start having trouble when you stand up. One episode is so bad you lose consciousness for twenty to thirty minutes. At the same time, you develop a horrific skin condition. A beautiful teenager who must cover her skin and face because of lesions and peeling skin! What are the most promising areas of research for skin conditions?

Moderating the immune response.

I want you to consider how unscientific the attack is on researchers like Frank and me as we are attempting to fix something that has gone terribly wrong in medicine. Frank and I have dedicated our lives to studying the immune system and its effect on human health.

We pioneered the field of modifying the function of the immune system through immunotherapies as a game-changing improvement to public health. However, we need to understand what we're doing and make sure we monitor the effects.

Nobody is acknowledging that vaccines are immunotherapy on a massive scale, and we have no idea what effect they are having on the population.

Should it come as any surprise that the bureaucrats of the Vaccine Court weren't interested in hearing our opinion?

In fact, they didn't even want to acknowledge our academic credentials.

* * *

The hammer was dropped on us by Special Master of the Vaccine Court Christian J. Moran on May 25, 2018, in the case of *Dominguez v. Secretary of Health and Human Services*, No. 12-378V. The issue was the hourly pay rate Frank and I were charging as PhD immunotherapy experts.

In our letter from October of 2017 to the attorney in charge of this case, we wrote about the amount of time Frank and I spent on the case:

We previously submitted a report in the case of G. D. vs. HHS. For the preparation of that report, we each first independently read and reviewed the clinical histories, previous annual physician visits, all diagnoses and treatments (exhibits 1–16).

We next spent more than 40 hours (each) searching and reviewing literature concerned with the development of vasculitis and granulomatous disease including extensive new literature regarding the innate and adaptive immune responses to recombinant vaccine antigens, attenuated viruses, vaccine excipients including but not limited to: aluminum, mercury, cellular debris, replication competent retroviruses, and virus-like particles and their effects on their effects on the immune system . . .

This rigorous review was accomplished because we were instructed to consider all possible causes of the development of vasculitis and granulomatous disease. Wherever possible we quoted directly from the records in preparing the summary of pertinent medical facts. We did not interpret any of the medical facts presented in Exhibits 1–16. We simply reported them. Wherever possible we quoted directly as written by the treating physician with the inclusion of the diagnostic codes used by the treating physicians.[14]

What was becoming clear to Frank and me was that the VICP judges didn't like us in their courts. Perhaps some of this friction was inevitable, as we were researchers, asking difficult and sometimes provocative questions, and the court itself was set up to be rather conservative.

I experienced a great deal of frustration because it seems to me a courtroom, or anything that is supposed to vaguely resemble one, should be the vehicle for a relentless search for truth. One cannot settle for incomplete answers.

Do any of us rest easy when an innocent person is sentenced to death? No, it disturbs all of us.

As I worked in the court, it appeared to me the pharmaceutical companies had pulled off a terrible trick, getting the United States government to be their shield against claims of consumer harm. Instead of warriors for public health, we had to settle for lazy bureaucrats simply interested in doing their time until their pensions vested.

In this case, we provided an invoice for $33,950.00 with a billing rate of $350 an hour. Pay attention to the disrespect of Special Master Christian Moran in his first discussion of his decision:

The balance of costs comes from an invoice of $33,950.00 for expert services provided by Ms. Mikovits and Mr. Ruscetti. Although both Ms. Mikovits

and Mr. Rusectti signed the expert reports, Ms. Mikovits was to be the sole
testifying witness and the analysis here thus refers solely to her.[15]

Is it possible to be any more demeaning in a few short sentences? Unless
Special Master Christian Moran has some new academic oversight of which
I'm unaware, Frank and I should be referred to as Dr. Ruscetti and Dr.
Mikovits. Last time I checked, we still had our PhDs.

And why is it that Special Master Moran seeks to completely discount
Frank's work on the report? Isn't one of the benefits of a report from two
researchers the idea that they check each other's work? Fine, I understand
that I've been attacked, and my reputation has been dragged through the
mud. But Frank's reputation is still as pure as the driven snow. Doesn't it
mean something that he did the work and signed the report, as well?

It gets worse:

> An hourly rate of $350 is consistent with the range of rates provided to expert
> medically-trained immunologists with extensive research experience that
> testify in the Vaccine Program. . . . It is true that Ms. Mikovits has been
> awarded a Ph.D. in biochemistry and neither Dr. Bellanti nor Dr. Shoenfeld
> has a Ph.D. While earning a Ph.D. is, itself, an accomplishment, an advanced
> degree is neither sufficient nor necessary for demonstrating the scientific
> expertise expected of expert witnesses in the Vaccine Program.[16]

I think it would probably have been a little less demeaning if he'd patted
me on the head, called me a good little girl, and sent me on my way with
a lollipop. This is the final insult from Special Master Moran to the report
Frank and I wrote:

> Based on her reputation and bona fides, Ms. Mikovits' credentials are simply
> not in the same league as experts who are paid $250 (or more) per hour. While
> this does not mean that Ms. Mikovits is incapable of providing expert testi-
> mony on specific topics, it does mean that she cannot expect to be paid the same
> hourly rate as those with much better reputations than she. Individuals with
> better reputations are, presumably, in far higher demand. Accordingly, based
> on the rate that the undersigned found reasonable for non-medically trained
> immunologists, $250 per hour, the undersigned makes an additional deduction
> of 40%. This deduction reflects Ms. Mikovits' relative lack of reputability in the
> field compared to comparable experts. This results in a rate of $150 per hour for
> a non-medically trained immunologist of Ms. Mikovits' reputation.[17]

Let's make sure we understand everything Special Master Moran is saying. Apparently, my twenty years of government research experience, including directing the world-renowned Lab of Anti-Viral Drug Mechanisms at the National Cancer Institute, mean next to nothing. The same could be said of my research, which changed the treatment of HIV-AIDS, saving the lives of millions. Special Master Moran also trashes the reputation of Frank, one of the greatest scientists to ever work in the field of cancer and who cofounded the discipline of human retrovirology. What does Special Master Moran make of the fact that Frank also signed the report in this case? Apparently nothing. Does Special Master Moran believe I spiked Frank's morning coffee as we were working together, then in his dazed state I got him to sign the report?

Let me tell you what Frank thinks about these shenanigans by Special Master Moran. He thinks that destroying the messengers is an old, cheap trick used to destroy the message. His number one rule has always been honesty and integrity. After that you are careful in trying to avoid errors, eliminating bias, and you are open in sharing data, resources, criticism, and ideas. He is frustrated and alarmed for the future by what he sees happening to projects after they are completed in a lab. He believes scientific communication is being distorted by the government regulatory agencies, the scientific journals, the mainstream media, and blogs by scientists promoting their own self-interest.

Frank and I are careful readers of scientific articles because we do actual research. One of Frank's favorite articles on this point was written by the longtime editor of the *British Medical Journal*, Richard Smith. The provocative title by this longtime journal editor is "Doctors Are Not Scientists." These are points Frank often shares with me:

> Some doctors are scientists—just as some politicians are scientists—but most are not.
>
> As medical students they were filled with information on biochemistry, anatomy, physiology, and other sciences, but information does not a scientist make—otherwise you could become a scientist by watching the Discovery channel. A scientist is somebody who constantly questions, generates falsifiable hypotheses, and collects data from well-designed experiments . . .
>
> The inevitable consequence is that most readers of most medical journals don't read the original articles. They may scan the abstract, but it's the rarest of beasts who reads an article from beginning to end, critically appraising it as he or she goes. Indeed, most doctors are incapable of critically appraising

an article. They have never been trained to do so. Instead, they must accept the judgment of the editorial team and its peer reviewers—until one of the rare beasts writes in and points out that a study is scientifically nonsensical.[18]

In most research projects, several different explanations can account for the results, equally well. Continued experimentation to reach a conclusion is the way science used to work, not through media pronouncements, government policies, and forced retractions.[19]

But it seems that the disrespect with which the Vaccine Court treats those who claim vaccines are causing damage is just the same as with the experts on their own side, who when they do their own research come to the same conclusion.

Witness the story of Dr. Andrew Zimmerman, the government's own expert witness in the Autism Omnibus Hearing before the vaccine court.

* * *

Okay, so apparently even though I have a PhD in biochemistry, have published more than fifty peer-reviewed articles, changed the treatment of HIV-AIDS impacting the lives of millions, and headed up a lab at the National Cancer Institute, I'm not qualified to be paid the same as other experts in Vaccine Court.

Somebody who could certainly demand more without the slightest bit of controversy is Sharyl Attkisson, the Emmy Award-winning former CBS News reporter who now has her own news show on the Sinclair Broadcasting Network called *Full Measure*. Attkisson is one of the few journalists who, in my opinion, has consistently covered the vaccine-autism issue, including most notably an interview with the former head of Health and Human Services, Dr. Bernadine Healy, who expressed strong support for looking at vulnerable subgroups of children who might have a negative reaction to a vaccination.

However, in January of 2019, Attkisson revealed what was arguably her biggest scoop on the vaccine-autism controversy when she did a segment on the allegations of the government's chief medical witness in the Autism-Omnibus hearings in 2007. This is from an article Attkisson wrote for *The Hill* in January of 2019 on the story:

> A world-renowned pro-vaccine medical expert is the newest voice adding to the body of evidence suggesting that vaccines can cause autism in certain susceptible children.

Pediatric neurologist Dr. Andrew Zimmerman originally served as the expert medical witness for the government, which defends vaccines in federal vaccine court. He had testified that vaccines do not cause autism in specific patients.

Dr. Zimmerman has now signed a bombshell sworn affidavit. He says that, during a group of 5,000 vaccine-autism cases being heard in court on June 15, 2007, he took aside the Department of Justice (DOJ) lawyers he worked for defending vaccines and told then he'd discovered "exceptions in which vaccinations could cause autism."

"I explained that in a subset of children, vaccine-induced fever and immune stimulation did cause regressive brain disease with features of autism spectrum disorder," Dr. Zimmerman now states. He said his opinion was based on "scientific advances" as well as his own experience with patients.[20]

This was the government's main scientific witness in what was the biggest controversy in medicine saying, "Yes, in certain circumstances, vaccines can cause autism." How was this not the biggest headline in public health of the twenty-first century?"

Dr. Zimmerman goes on to say that once the DOJ lawyers learned of his position, they quickly fired him as an expert witness and kept his opinion secret from other parents and the rest of the public.

What's worse, he says the DOJ went on to misrepresent his opinion in federal vaccine court to continue to debunk vaccine-autism claims.

Records show that on June 18, 2007, a DOJ attorney to whom Dr. Zimmerman spoke told the vaccine court: "We know [Dr. Zimmerman's] views on this issue . . . There is no scientific basis for a connection" between vaccines and autism.

Dr. Zimmerman now calls that "highly misleading" and says he'd told them the opposite.[21]

According to Attkisson, Dr. Zimmerman knew in June of 2007 that vaccines could cause autism in certain susceptible children, especially among those who developed a fever after a vaccination. I'd like to point out that the government's own expert witness came to this conclusion a full two years before I even broached such a possibility. How can he call himself a doctor if he knowingly avoids the Hippocratic Oath for more than a decade, which states, "First, do no harm"?

I'm really struggling to contain my anger as I write these words.

Just for the sake of argument, let's say Zimmerman attempted to do the right thing by telling the government lawyers that vaccines could cause autism in a certain subset of children.

For this act of honesty, he gets kicked out of Vaccine Court. Fine, I respect that.

And then he does nothing for the next ten years?

No call to the *New York Times* or the *Washington Post*? In all those years, he could easily have picked up the phone to dial one of these papers and said, "Hey, Mr. Reporter, I'm the director of Medical Research for Autism at the Kennedy-Krieger Institute for Johns Hopkins University, and I think I know what might be causing at least some of the autism in the United States? Are you interested in interviewing me?"

Was he unaware of how the country and families are being ripped apart over this question? Did he have any responsibility to speak up? This was a man who understood what was going on years before I did, yet I'm the one who spoke up and got my reputation dragged through the mud. I voiced my concerns, was arrested in a show of force more suited for some drug lord, jailed for five days without bail, had my mug shot taken and put in *Science*, lost my career, went bankrupt, and now I'm insulted and refused rightful pay for hundreds of hours of honest work by a moronic corrupt special master in Vaccine Court?

Maybe I'm doing all the talking for Dr. Andrew Zimmerman.

Sharyl Attkisson offered Dr. Zimmerman the opportunity to be interviewed on camera, but he declined, sending her instead to his signed affidavit. So, don't let me speak for Dr. Zimmerman, or Sharyl Attkisson, even though she's an Emmy-winning journalist.

This is Andrew Zimmerman in his own words:

I, Andrew Walter Zimmerman, M. D., do hereby state under oath as follows:

> I am a board certified, pediatric neurologist and former director of Medical Research, Center for Autism and Related Disorders, Kennedy Krieger Institute, and Johns Hopkins University School of Medicine.
>
> I was a reviewer for the National Academy of Sciences 2004 report entitled IMMUNIZATION SAFETY REVIEW: VACCINES AND AUTISM, which was prepared by the Immunization Safety Review Committee, at the request of the Centers for Disease Control and Prevention (CDC), the National Institutes of Health (NIH), and the Institute of Medicine (IOM) . . .[22]

The next portion of Zimmerman's affidavit relates to another autism case in which he did not believe vaccines contributed to that child's autism. Then Zimmerman lays out in his own words exactly what he told the Department of Justice lawyers and their response to what Zimmerman was planning to say in his testimony:

> On Friday, June 15th, 2007, I was present during a portion of the O.A.P. [Omnibus Autism Proceeding] to hear the testimony of the Petitioner's expert in the field of pediatric neurology, Dr. Marcel Kinsbourne. During a break in the proceedings, I spoke with DOJ attorneys and specifically the lead DOJ attorney, Vincent Matanoski in order to clarify my written expert opinion.
>
> I clarified that my written expert opinion regarding Michelle Cedillo was a case specific opinion as to Michelle Cedillo. My written expert opinion regarding Michelle Cedillo was not intended to be a blanket statement as to all children and all medical science.
>
> I explained that I was of the opinion that there were exceptions in which vaccinations could cause autism.[23]

My coauthor, Kent, an attorney, loves to quote from legal documents. He likes to have the exhibits shown to you in their entirety. But I prefer to break them up into more easily digestible pieces and summarize. Zimmerman was the government's expert on vaccines and gave an opinion in the case of Michelle Cedillo. Based on his review of her records, he did not believe vaccines were responsible for her autism.

However, he did not intend his opinion about Cedillo be a stand-in for all claims of vaccine injury resulting in autism. In fact, he believed there were situations in which vaccines COULD cause autism.

Let that sink in.

Imagine if the government said we're reviewing claims of alien abduction. In the case of one person who claimed to have been abducted, the government comes out and says we don't believe the evidence supports that claim. But what if the government came out and said, "Hey, this other guy, we believe he was abducted by aliens. And just to add a little more context, we think 30 percent of the people who claim abduction by aliens are telling the truth."

The entire world would change in that single moment.

Vaccines are harming a certain group of children as they go for their pediatric wellness visits, the government is concealing this information from the doctors, and the children who are affected are damaged for life.

And the government lawyers, and probably their superiors, knew this since at least 2007.

In his affidavit, Zimmerman further detailed the basis for his belief that vaccines could cause autism in some children:

> More specifically, I explained that in a subset of children with an under-lying mitochondrial dysfunction, vaccine induced fever and immune stim-ulation that exceeded metabolic energy reserves could, and in at least one of my patients, did cause regressive encephalopathy with features of autism spectrum disorder.
>
> I explained that my opinion regarding exceptions in which vaccines could cause autism was based on advances in science, medicine, and clinical research of one of my patients in particular.
>
> For confidentiality reasons, I did not state the name of my patient. However, I specifically referenced and discussed with Mr. Matanoski and the other DOJ attorneys that were present, the medical paper, **Developmental Regression and Mitochondrial Dysfunction in a Child with Autism**, which was published in the *Journal of Child Neurology* and co-authored by Jon Poling, M.D., Ph.D., Richard Frye, M.D., Ph.D., John Shoffner, M.D., and Andrew W. Zimmerman, M.D., a copy of which is attached as exhibit C.
>
> Shortly after I clarified my opinions with the DOJ attorneys, I was con-tacted by one of the junior DOJ attorneys and informed that I would no lon-ger be needed as an expert witness on behalf of H.H.S. [Health and Human Services] The telephone call in which I was informed that the DOJ would no longer need me as a witness on behalf of H.H.S. occurred after the above referenced conversation on Friday, June 15, 2007, and before Monday, June 18, 2007.
>
> To the best of my recollection, I was scheduled to testify on behalf of H.H.S. on Monday, June 18, 2007.[24]

Zimmerman believes a vaccination and the accompanying immune stimu-lation might overwhelm a child's energy supply and cause an encephalopa-thy (brain swelling) resulting in "features" of autism. I have a little trouble with Zimmerman's phrasing of "features of autism spectrum disorder." It means autism, and he should've just said it. There is no difference between "autism" and "features of autism spectrum disorder."

Then we come to one of the oldest stories in the world: the cover-up of an embarrassing truth.

The government, in preparing its case, looked for and found one of the best experts in the country, Dr. Andrew Zimmerman of Johns Hopkins University.

An expert like Zimmerman would tend to be more conservative in his opinions, likely to discount newly expressed and intriguing theories until a robust set of evidence has been presented. To some this may be a drawback, but to others it would be a benefit. If Zimmerman believes something to be true, all sides can be comfortable his view is well supported, if not invulnerable.

Zimmerman gave his opinion in the Cedillo case that her autism was not caused by vaccines.

But he did tell the government attorneys that his medical opinion was that vaccines could cause autism in certain children, based on a preexisting weakness in their energy production. He had observed at least one child who fit this diagnosis and published about it.

Zimmerman told the DOJ attorneys of this opinion on Friday, June 15, 2007, knowing he was scheduled to testify in another case the following Monday.

Sometime during that weekend, Zimmerman was called by the Department of Justice and told his services would no longer be needed.

Yes, I know, correlation does not mean causation.

Just because when I walk into a crosswalk and go flying thirty feet at the same time a car barrels through the same crosswalk doesn't mean I was hit by that car.

Maybe there's some other explanation.

Like aliens did it.

And at the same time, they smashed up the car with their laser beam.

Zimmerman continues in his affidavit with information about the child he observed, Hannah Poling, who happened to be the daughter of one of Zimmerman's fellow neurologist colleagues, Dr. Jon Poling:

> At the time of the above referenced conversation with the DOJ, I did not know that Hazlehurst v. HHS or Poling v. HHS were potential cases in the OAP. [Omnibus Autism Proceeding]
>
> It is my understanding that the HHS concession in Poling v. H.H.S. has become common knowledge and has been published by international media. Among other news coverage, I reviewed the CNN interview in which Dr. Julie Gerberding, the former head of the CDC discussed the concession by H.H.S. in Poling v. H.H.S. and the interview with Dr. Jon Poling, the father of the child whose case was conceded.

The summary language, "the vaccination . . . significantly aggravated an underlying mitochondrial disorder, which predisposed her to deficits in cellular energy metabolism, and manifested as a regressive encephalopathy with features of autism spectrum disorder" is in essence the chain of causation that I explained to the DOJ attorneys including Vincent Matanoski during the above referenced conversation on June 15, 2007.

I have reviewed extensive genetic, metabolic and other medical records of William "Yates" Hazlehurst. In my opinion, and to a reasonable degree of medical certainty, Yates Hazlehurst suffered regressive encephalopathy with features of autism spectrum disorder as a result of a vaccine injury in the same manner as described in the DOJ concession in Poling v H.H.S., with the additional factors that Yates Hazlehurst was vaccinated while ill, administered antibiotics and after previously suffering from symptoms consistent with a severe adverse reaction.[25]

Those paragraphs may be the most important of the affidavit, but I'll need to provide a little background for you to fully understand their importance.

The setup of the Omnibus Autism Proceeding was that the five test cases would stand in for the more than five thousand claims that had been filed. Poling was supposed to have been the first, but it was settled prior to the start of the proceeding.

Cedillo was next, and as you've read, the evidence in that case was not strong enough in Zimmerman's opinion to support a finding that vaccines had caused Michelle Cedillo's autism.

(My coauthor, Kent, strongly disagrees with that conclusion, but let's leave that aside for the moment.)

The case of William "Yates" Hazlehurst was scheduled to be the third case, but since Poling had settled (and was subject to a nondisclosure agreement), it would be the next case after Cedillo.

In Zimmerman's opinion, Hazlehurst's closely resembled the Hannah Poling case, meaning that vaccines did cause his injuries, and he was prepared to testify to that fact. This is what Dr. Zimmerman wrote in a two-page letter to the Polings' attorney on November 30, 2007, about Hannah's case:

The cause for regressive encephalopathy in Hannah at age 19 months was underlying mitochondrial dysfunction, exacerbated by vaccine-induced fever and immune stimulation that exceeded metabolic energy reserves. This acute expenditure of metabolic reserves led to irreversible brain injury.[26]

How large of a problem would a decision in favor of Yates Hazlehurst have been to the United States government?

One of Dr. Zimmerman's colleagues in the Omnibus Autism Proceeding was Dr. Richard Kelley, a professor of pediatrics from Johns Hopkins University (Kennedy-Krieger Institute), who is acknowledged to be one of the country's leading experts on mitochondrial dysfunction. In 2018, Dr. Zimmerman and Dr. Kelley joined a lawsuit on behalf of Yates Hazlehurst and have given depositions as to their opinion regarding what happened to Yates and the events surrounding the Omnibus Autism Proceeding. This is what Dr. Kelley said in his written affidavit for the Hazlehurst case:

> I also find with a high degree of medical certainty, that the set of immunizations administered to Yates at age 11 months while he was ill was the immediate cause of his autistic regression because of the effect of these immunizations to further impair the ability of his weakened mitochondria to supply adequate amounts of energy for the brain, the highest energy consuming tissue in the body.[27]

Let's use that as something of a working hypothesis for autism. For some reason, the mitochondria are not working at an optimal level. We know that retroviruses tend to lower the activity of the mitochondria, so if we had many children with active retroviruses, the effect would be lowered energy. I think that's a reasonable hypothesis.

J.B. Handley, in his excellent 2018 book, *How to End the Autism Epidemic*, quotes extensively from the Dr. Kelley deposition in the Hazlehurst case. Here is an excerpt from the deposition:

> **Lawyer:** Would you say that you are an expert in mitochondrial dysfunction but not in autism. Would that be a fair way to describe it?
>
> **Dr. Kelley:** I am an expert in mitochondrial disease. And I am an expert in the aspect of autism that pertain to the roughly 25, 30, 40 percent of children who have autism based on mitochondrial dysfunction.[28]

Let's do the math.

In 2007 it was estimated there were close to a million children with autism in the United States. Kelley believed in 2007 that roughly one-third or more of the children with autism have a mitochondrial dysfunction that is responsible for their autism. Vaccines will drain mitochondrial reserves.

Let's put that number at three hundred thousand children (30 percent of a million children), a conservative estimate, as I'm sure Andrew Zimmerman would approve.

It's been estimated that the lifetime care cost for a child with autism is three million dollars.

What is three hundred thousand multiplied by three million?

It's nine hundred billion dollars.

Nine hundred billion dollars is a conservative estimate of damages for which the federal government would be responsible if Vincent Matanoski had allowed the government's own medical witness to testify in the Yates Hazlehurst case on June 18, 2007.

But it gets even worse, if such a thing were possible.

It wasn't that Department of Justice lawyer Matanoski excluded the evidence provided by Dr. Zimmerman; he actively misrepresented what Dr. Zimmerman had specifically told the Department of Justice lawyers.

Zimmerman's statement continues:

I have reviewed the attached portions of the transcript, of Vincent Matanoski's closing argument on Hazlehurst v. H.H.S., which is attached as exhibit D. The relevant portion of the transcript states as follows:

> I did want to mention one thing about an expert, who did not appear here, but his name has been mentioned several times, and that was Dr. Zimmerman.
>
> Dr. Zimmerman actually has not appeared here, but he has given evidence on this issue and it has appeared in the Cedillo case. I just wanted to read briefly because his name was mentioned several times by Petitioners in this matter. What his views were on these theories, and I'm going to quote from Respondent's Exhibit FF in the Cedillo case, which is part of the record in this case as I understand it.
>
> "There is no scientific basis for a connection between measles, mumps and rubella MMR vaccine or mercury intoxication in autism despite well-intentioned and thoughtful hypotheses and widespread beliefs about apparent connection with autism and regression. There's no sound evidence to support a causative relationship with exposure to both or either MMR and/or mercury."
>
> We know his views on this issue.
>
> In my opinion, the statement by Mr. Matanoski during his closing argument regarding my opinion was highly misleading and not an accurate reflection of my opinion out of context. My opinion as to Michelle Cedillo was case-specific. I was only referring to the medical evidence I had reviewed regarding her. My opinion regarding Michelle Cedillo was not intended to

be a blanket statement as to all children and all medical science. Second, as explained above, I specifically explained to Mr. Matanoski and the other DOJ attorneys who were present that there were exceptions in which vaccinations could cause autism.

In my opinion, it was highly misleading for the Department of Justice to continue to use my original written expert opinion, as to Michelle Cedillo, as evidence against the remaining petitions in the O.A.P. in light of the above referenced information which I explained to the DOJ attorneys while omitting the caveat regarding exceptions in which vaccinations could cause autism.[29]

The mind boggles when it reads a statement such as that from Dr. Zimmerman. What Matanoski is alleged to have done is to misrepresent evidence that vaccine injuries were taking place, even when verified by their own medical experts.

I have come to understand why parents of vaccine-injured children view the federal government as their enemy.

* * *

In retrospect, I should perhaps feel fortunate that the Vaccine Court published a document in which they refused to give Frank and me our proper professional designations in 2018.

At least I have their disrespect crimes in writing.

Zimmerman just got a phone call in 2007.

We were saying the same things in 2018 that Zimmerman was saying in 2007, with the benefit of eleven years of additional scientific research.

And what was a nine-hundred-billion-dollar liability in 2007 has only grown larger, with the estimate now of 1.8 million children in the United States having autism.

My identity may have been stolen in Vaccine Court, but the corruption in that court has robbed children and families in America of so much more.

Justice must come for all. I pray the day of reckoning is close.

CHAPTER NINE

What I Really Think about HIV and Ebola

It's my intention to make this the simplest chapter of the book to understand.

The downside is I'm likely to completely destroy your faith in vaccines as well as a big chunk of public health.

When this chapter is over, I don't think you'll ever look at vaccines in the same way. I apologize in advance for any problems this may cause among your friends and family.

I strongly encourage you to check my work to see if I've made any errors of fact or interpretation, but I feel as strongly about this claim as anything in my life.

* * *

The first concept I want you to understand is called zoonosis, or more properly, zoonotic diseases. If you think it's a funny word and you immediately imagined a picture of a zoo, you wouldn't be far off. Zoonosis literally means a disease which can spread between animals and humans.

It may surprise you given my reputation as a renegade, but I believe the Centers for Disease Control (CDC) can occasionally provide good, basic information to the public. Here is what they have on their website about zoonotic diseases.

> Every year, tens of thousands of Americans will get sick from diseases spread between animals and people. These are known as zoonotic diseases.

Zoonotic means infectious diseases that are spread between animals and people. Because these diseases cause sickness or death in people, CDC is always tracking them.

Animals provide many benefits to people. Many people interact with animals in their daily lives, both at home and away from home. Pets offer companionship and entertainment, with millions of households having one or more pets . . .

However, some animals can carry harmful genes that can be shared with people and cause illness—these are known as zoonotic diseases or zoonoses. Zoonotic diseases are caused by harmful germs like viruses, bacteria, parasites, and fungi. These germs can cause many different types of illnesses in people and animals ranging from mild to serious illness to death. Some animals can appear healthy even when they are carrying germs that can make people sick.

Zoonotic diseases are very common, both in the United States and around the world. Scientists estimate that more than 6 out of every 10 known infectious diseases in people are spread from animals, and 3 out of every 4 new emerging infectious diseases in people are spread from animals.[1]

That's good solid information from the CDC and underscores the scope of the problem. Tens of thousands of Americans at the very least will get sick every year from illnesses spread by some form of association with animals. Zoonotic diseases account for more than 60 percent of known infectious diseases and at least 75 percent of new emerging infectious diseases. This is probably one of the biggest challenges in public heath you've never heard about, even though you can find it prominently displayed on the CDC website.

What are some of these diseases, you might ask? This is a partial list: anthrax, bird flu, bovine tuberculosis, cat scratch fever, dengue fever, Ebola, encephalitis from ticks, enzootic abortion, hemorrhagic colitis, hepatitis E, listeria infection, Lyme disease, malaria, parrot fever, plague, rabies, rat-bite fever, ringworm, Rocky Mountain spotted fever, swine flu, toxoplasmosis, West Nile virus, and zoonotic diphtheria.[2]

And the biggest zoonotic disease that is not covered in that list is HIV-AIDS, which affected more than sixty million people leading to our world's greatest modern plague. While a good deal of ink has been dedicated to the question of how prejudice against the gay lifestyle delayed efforts to properly address the disease, our job as scientists is to provide an explanation of events in the past and how problems might be avoided in the future.

Let's make sure we understand our terms. The human immunodeficiency virus (HIV) is linked to the condition known as acquired immunodeficiency syndrome (AIDS). There was a brief time when the disease was known as gay-related immune disease (GRID), and many activists claim the name change to AIDS prompted a more balanced examination of the disease. Perhaps this is true.

But what of the genesis of the retrovirus HIV? Where did it come from?

We have a clear answer.

It came from a primate. The field agrees the precursor to HIV was the simian immunodeficiency virus or SIV.

To be more precise, the human virus came from a monkey or chimpanzee virus.

After that, any potential areas of agreement break down.

I vividly recall working as a technician at the Biological Response Modifiers Program at Fort Detrick in the early 1980s, where it was my job to isolate HIV from samples and find a cell line or tissue to grow the virus. If you can't grow the virus outside a human body, you can't study it.

What we were told at the time was that the disease probably jumped to humans as a result of Africans forgetting how to cook their food, in this case chimpanzees, often referred to as "bush meat," and that the promiscuous sexual lifestyles of African peoples (with implications of possible bestiality with primates) led to the cross-species jump and spread of the virus among the human population. I am now appalled that at the time we did not more closely question these assumptions.

Since that time, there have been two competing and, in my mind, closely related theories of how a chimpanzee retrovirus made the jump to humans.

The first was popularized by the journalist Edward Hooper and expanded upon in his lengthy 1999 book, *The River: A Journey to the Source of HIV and AIDS*, for which he conducted more than six hundred interviews. Of the interviews conducted, Hooper was most impressed with evolutionary biologist Bill Hamilton, who, along with other independent researchers such as Louis Pascal, Tom Curtis, and Blaine Elswood, was proposing an idea that, before my work with XMRV, I would have found quite radical. They proposed that the jump from chimpanzees to humans came as a result of vaccine trials in the Belgian Congo from 1957 to 1960 in which more than five hundred common chimpanzees and bonobos (pygmy chimpanzees) were killed so that their kidney cells and sera could be used to grow the oral polio vaccine. This vaccine was subsequently administered to more than a million Africans during that time period.

Hooper suggests there was great resistance to this idea, since even in the late 1950s and early 1960s there was little public support for the idea of using chimpanzees in such medical experiments. In addition, the Belgian royal family was publicly supporting the idea of wildlife conservation, and the revelation of these actions would go against that image. Hooper believes another reason for resistance to his idea is that if his theory was accepted, it would shake public confidence in the medical establishment as well as lead to multibillion-dollar class action lawsuits for the AIDS epidemic.

This is what Edward Hooper has written on his website about the circumstances surrounding the creation of HIV-AIDS from these experiments and why it makes more sense than the competing bush-meat and "cut hunter" theory.

> By contrast, the oral polio vaccine (OPV) theory proposes that an experimental OPV that had been locally prepared in chimpanzee cells and administered by mouth, or "fed," to nearly one million Africans in vaccine trials staged in the then Belgian-ruled territories of the Belgian Congo and Ruanda-Urundi between 1957 and 1960, represents the origin of the AIDS pandemic. It provides a historically-supported background: that between 1956 and 1959 over 500 common chimpanzees (Pan troglodytes schweinfurthi and Pan troglodytes troglodytes) and bonobos or pygmy chimpanzees (Pan paniscus) were housed together at Lindi Camp (Near Stanleyville in the Belgian Congo, or DRC).[3]

Hooper goes on to explain that the use of chimpanzees was not technically prohibited by any conventions, but that in most countries around the world at the time, Asian macaques were used in polio virus preparation. As for the claim of other outbreaks, Hooper believes they fit comfortably within the bounds of his theory:

> The OPV theory ascribes the minor outbreaks of AIDS caused by other variants of HIV-1 (Group O, Group N and the more controversial "Group P") to other polio vaccines (both oral and injected) that were prepared in the cells of chimpanzees and administered in French Equatorial Africa (including Congo Brazzaville and Gabon) in the same late fifties period. It ascribes the outbreak of AIDS from HIV-2 (of which it maintains that only two were epidemic outbreaks) to other polio vaccines (both oral and injected) that were prepared in the cells of sooty mangabeys (or other monkeys that had been caged with sooty mangabeys) and administered in French West Africa in 1956–1960. All

the other HIV-2 groups that are claimed by bush-meat theorists have infected just a single person, and some OPV theory supporters argue that dead-end, non-transmissible infections such as these are the natural fate of SIVs that infect human beings via the bushmeat route: that unless they are introduced in an artificial manner (as via a vaccine), they simply die out.[4]

I find myself in complete agreement with Hooper's analysis. Yes, viruses can jump from one species to another, possibly by the eating of an infected animal. But the natural process of degradation by the digestive system, as well as cooking (even when poorly done) is likely to inactivate most pathogens.

The second plausible theory, which has come to be known as the "bush-meat" theory, is that sometime early in the twentieth century an individual became infected with SIV from handling chimpanzees or chimpanzee bush-meat. It usually involves the actions of some anonymous native hunter (often called the "cut hunter" because he cut himself shortly after having killed a chimpanzee). Added to this scenario is urbanization encroaching upon the jungle, allowing it to be spread by those newly introduced western evils of prostitution and intravenous drug usage.

This theory has recently been expanded by the science writer David Quammen in *The Chimp and the River: How AIDS Emerged from an African Forest*. Based on an extrapolation of some scientific data, Quammen sets the date for this viral transfer from chimp to human around 1908 in the area known as Leopoldville (later Kinshasa) in the Democratic Republic of the Congo. The description is vivid, and plausible, but I question much about it:

> Let's give him due stature: not just a cut hunter but the Cut Hunter. Assuming he lived hereabouts in the first decade of the twentieth century, he probably captured his chimpanzee with a snare made from a forest vine, or in some other form of a trap, and then killed the animal with a spear. He may have been a Baka Pygmy man, living independently with his extended family in the forest or functioning as sort of a serf under the "protection" of a Bantu village chief . . . There's no way of establishing his identity, nor even his ethnicity, but this remote southeastern corner of what was then Germany's Kamerun colony offered plenty of candidates . . .
>
> The chimp too, tethered by a foot or a hand, would have been terrified as the man approached, but also angry and strong and dangerous. Maybe the man killed it without getting hurt; if so, he was lucky. Maybe there was an ugly fight; he might even have been pummeled by the chimp, or badly bitten. But he won. Then he would have butchered his prey, probably on the spot . . .

I imagine him opening a long, sudden slice across the back of his left hand, into the muscular web between thumb and forefinger, his flesh smiling out pink and raw almost before he saw the damage or felt it, because his blade was so sharp . . . His blood flowed out and mingled with the chimp's, the chimp's flowed in and mingled with his, so that he couldn't quite tell which was which. He was up to his elbows in gore. He wiped his hand. Blood leaked again into his cut, dribbled again into it from the chimp, and again he wiped. He had no way of knowing, no language or words or thoughts by which to conceive, that this animal was SIV-positive. The idea didn't exist in 1908.[5]

It's a possible scenario. I can't say something like that didn't happen. I just wonder why it hadn't happened many centuries earlier. Africans had been hunting chimpanzees for thousands of years and cooking them. An interesting addition Quammen makes to the theory is that subsequently the virus spreads slowly among the population. But then starting in 1917, there were vaccination campaigns against sleeping sickness by European doctors who used glass syringes that were reusable. One French colonial doctor during a two-year period treated more than five thousand cases with only six syringes.[6] These campaigns peaked in the early 1950s, and by that time the precursor to the deadly strain of HIV had arisen.

When you think about it, you come to the realization there's a battle of narratives, with science having a definite preference for one over the other.

In the first scenario, unwitting scientists release a plague of massive proportions on the population because of their use of questionable animal experiments, infecting more than sixty million people and causing the death of at least thirty-nine million.

In the second scenario, a chance event in a jungle encounter with an infected chimpanzee leads to cross-species transmission, then because it's always a good play to blame urbanization and prostitution, as well as maybe a little inadvertent help from western medicine, and you have a new disease!

Is it any surprise that scientists far prefer scenario number two?

While I can't come to a definitive conclusion as to which scenario is more likely, the first one, in which chimpanzees are directly harvested for their organs and the growing of polio virus, makes the most sense to me. It doesn't have as many moving parts.

The virus is in a certain percentage of the five hundred chimpanzees sacrificed. They're cut up, then used to grow polio vaccine, which is then given orally to nearly a million Africans. And there's another part of the story that makes Hooper's account sound more plausible.

After Hooper made his claim that the oral polio vaccine grown in chimpanzee tissue and given to humans was the source of the HIV-AIDS epidemic, there was an "investigation." As I read Hooper's account, it sounded a lot like the Ian Lipkin investigation into XMRV.

In this great investigation, they sampled stocks of polio vaccine from the 1957–1960 period for traces of chimpanzee DNA, or simian virus. Lo and behold, they found NOTHING!

There's just one problem. All of the samples they used were from the United States.

They did not have any samples of polio vaccine from Africa. The samples of polio vaccine from the United States had never used chimpanzee tissues as a growth medium or cell line.

They DID NOT test African samples of the oral polio vaccine for that which used chimpanzee tissue.

This is what Hooper wrote about the supposed investigation into the use of chimpanzee tissue in the development of the polio vaccine that was performed by the Royal Society in September of 2000:

> Instead of the open and honest debate, and the even-handed investigation of the OPV theory, which had been promised, what actually took place was a carefully-planned attempt to suppress the theory by fair means or foul. The conference became focused around the testing of samples of CHAT vaccine which the parent institute (The Wistar in Philadelphia) had finally released for independent analysis. The vaccinators and the meeting organizers insisted that the vaccine samples were representative of the batches which had been prepared for use in Africa. Since they tested negative for HIV, SIV, and chimpanzee DNA, they concluded that the OPV hypothesis had been disproved— and an acquiescent press largely concurred.
>
> The reality, however, was very different. None of the tested samples had ever been near Africa, let alone prepared for the African trials.
>
> As the weeks and months passed after the meeting, it became clear that a carefully-organized whitewash was being carried out, partly by the original protagonists (who had, among other things, leant on witnesses to change key parts of their stories), and partly by well-meaning research scientists who could not countenance the possibility that their work of the last ten years might be erroneous, and who secondly were unwilling or unable to imagine that their peers might not be telling the truth.[7]

I've never met Edward Hooper, but he was writing this in 2004, years before I ever pursued a similar line of inquiry. I'm sure he must have undergone a similar evolution, from hopeful questioner to disillusioned critic. Is it so difficult to imagine that members of an organization will not believe the worst about their own members? Don't we see the same pattern among police, members of the clergy, and our political class? Isn't it the rare member of an organization who sees the flaws of their own group?

We leave scientists alone in their research and practice, expecting somehow that they can self-police. But we do not even allow the police to self-police. There are citizen review boards, internal affairs, and oversight committees.

In a similar manner, we have learned through bitter experience that just because a person is a member of the clergy doesn't mean they are not capable of crimes against children. We are also beginning to understand that these crimes are not just committed against the flock, but there are more nuns and sisters in the Catholic Church who are coming forward with stories of rape and sexual assault committed by the male members of the clergy. These stories sicken us, but perhaps we have simply trusted too much in unaccountable authorities.

Now, maybe Hooper is wrong about his accusation, but it sounds like a pretty serious charge to make. Given what I observed in the "official" investigation into XMRV and the way they cavalierly rewrote basic principles of virology, I'm more inclined to believe Hooper's account. The scientific establishment tells their stories and expects we will believe them.

The simian virus gets transferred into humans, and then the question becomes one of immune activation. What happened in the gay community in the late 1970s and early 1980s?

There was a great deal of recreational drug usage, and it's a scientific fact that anal sex, with its subsequent tearing of tissue, promotes immune activation. The sexual revolution for the heterosexual population, and the lesbian population as well, did not involve such risks.

* * *

Are dangers to the human population limited to the use of chimpanzee tissue in the development of medical products, or is it a more general question of any animal tissue? I tend to believe the latter and use as an example the controversy over Simian Virus-40 (SV-40) in the same antipolio campaign of the 1950s and 1960s. The concern is that these viruses may lie dormant in people until some form of immune stimulation, just as we saw with HIV and the gay lifestyle that included multiple partners, sexual activity that involved anal tearing, and high recreational drug usage.

In her Pulitzer Prize-nominated book, *The Pentagon's Brain*, detailing the work of the Defense Advanced Research Projects Agency (DARPA), author Annie Jacobsen provides a brief overview of this controversy. One of the DARPA scientists she interviewed for the book was microbiologist Stephen Block, and this is what she wrote:

> If the notion of a stealth virus, or silent load, sounded improbable, Block cited a little-known controversy involving the anti-polio vaccination campaign of the late 1950s and early 1960s. According to Block, during this effort millions of Americans risked contracting the "cryptic human infection" of monkey virus, without ever being told. "These vaccines," writes Block, "were prepared using live African green monkey kidney cells, and batches became contaminated by low levels of a monkey virus, Simian virus 40 (SV 40), which eluded the quality control procedures of the day. As a result, large numbers of people—probably millions in fact—were inadvertently exposed to SV 40."[8]

The controversy over SV 40 was whether it would ultimately lead to cancers in humans decades later. The virus would often be found in cancerous tissues, raising the question of whether it was simply a passenger or a causative agent. Again, this is the same concern raised by the finding of bovine leukemia virus in samples of breast cancer tissue and whether the use of growth hormone in the cattle was prompting the expression of this virus. It's also worth noting that essentially zero testing is done these days of animal viruses contaminating our vaccines or other medical products. Our medical authorities simply assume we're all tough enough to fight off these contaminating viruses. Jacobsen continues:

> Block says that two outcomes of this medical disaster remain debated. One side says the 98 million people vaccinated dodged a bullet. The other side believes there is evidence the vaccine did harm. "A great deal of speculation occurs about whether [simian virus] may be responsible for some diseases" that manifests much later in the vaccinated person's life, says Block, including cancer.[9]

Let's go to the doubters who agree that at least ninety-eight million people were inoculated with a polio virus that was contaminated with SV-40. Is it reasonable to assume that if you fire ninety-eight million bullets, none will cause any harm?

And this only considers the polio vaccine. Every vaccine has been grown in animal tissue, usually of several different species, including monkey, mouse, bird, and cow. Each one of these cross-species events has the potential to transfer a pathogen to humans, or to create some new strain that can cause harm. We have fired several billion bullets of biological ammunition at the human species, and it is the height of arrogance to believe we have caused zero damage.

* * *

In my discussion of Ebola, I want to highlight an idea many others have been discussing in recent years in one form or another, but that is likely to reconfigure how we go about promoting health.

For more than a century, we've been promoting two different concepts, which when you think about it are fundamentally at odds with each other. The first concept is that we need to ensure our food, air, and water is as free of pollutants and pathogens as possible. I have no problem saying I'm in favor of this effort a hundred percent.

The second concept involves the belief that we need to prime our immune system with weakened or dead pathogens in order to deal with any challenges that might come our way over the course of a lifetime. This is the fundamental idea behind vaccinations.

I think we're committing overkill of a good idea, and it is leading to tragic unintended consequences.

If we are living in a relatively clean environment, and we have proper nutrition, then our immune system is going to develop naturally. I've heard many activists make the claim that vaccinations create "fake" immunity, whereas a robust immune system is more likely to be able to respond to any pathogens that might be encountered in the course of our regular lives in a relatively clean environment. I have a great deal of sympathy for this position.

Here's one of the things that concerns me as an unintended outcome of vaccinations. If we are taking all the possible pathogens, serious and mild, that we might conceivably encounter over the course of a lifetime, we are being exposed to more pathogens than might otherwise be expected. That exposure is in most cases bypassing critical immunity, such as the skin and gut. Each challenge to the immune system by a vaccination has the potential to dysregulate the immune system. We have no idea what happens when multiple different pathogens are injected at the same time.

The paper that most catalyzed my thinking in this area was published in November of 2009, just around the same time I published my XMRV findings in *Science*. That paper was published by some of my former colleagues at the National Institutes of Health. The title of it was "War and Peace Between Microbes: HIV-1 Interactions with Co-infecting Viruses," and it may be one of the most important papers of the last twenty-five years.

From the introduction:

> The development of immunology in the last century led to the concept of a healthy "germ-free" human body that repulses and eliminates invading microbes by generating effective immune reactions. Through the years, it became clear that a healthy host is not germ free and does not always fight "germs" but may, rather, live in symbiotic relationships with some of them . . .
>
> Simian immunodeficiency virus (SIV), which has circulated in sooty mangabeys (SIVsm) and African green monkeys (SIVagm) for a long time, does not cause AIDS, despite high replication and lack of immune control (Paiardini et al., 2009). SIV began to infect chimpanzees (SIVcpz) more recently than sooty mangabeys or African green monkeys, and it causes a disease, which is apparently less severe than the one that human immunodeficiency virus type 1 (HIV-1) causes (Keele et al. 2009) . . . Its invasion greatly imbalances the body's equilibrium with other microbes . . .
>
> The uncontrolled replication of the symbiotic and newly invading microbes contributes to the imbalance of the immune system by perpetuating its uncoordinated activation, which, in turn, further accelerates progression toward AIDS (Figure 1B). Thus, like an orchestra that after a sudden disappearance of the conductor continues to play fragments of the scored melodies, in HIV-1-infected individuals, the immune system continues to play out a chaotic and ineffective attack against microbes.[10]

There's a good deal to unpack in all of this, but it's pretty straightforward.

We've come to understand that it's not enough to simply keep us away from germs, but that our immune system should be strong enough to either defeat those germs to which we're exposed or reach a state of equilibrium with them. We are finding that natural exposures, such as having the measles, not only provides lifelong immunity, but also tunes the system to make it less likely you'll have certain cancers when you're older.

I've already talked about the simian immunodeficiency virus (SIV) and the fact that it can be found in chimpanzees, sooty mangabeys, and African green monkeys. Viruses and microbes have the capacity to unbalance the immune system. This can lead to activation of previously dormant viruses, or infection by others that have now turned virulent due to decreased immune function.

The image of an orchestra that has suddenly lost its conductor is one of the best metaphors I've ever read for understanding this issue. Without the conductor, the orchestra can continue playing music, but it's likely to be the wrong song.

And what's at stake is our health.

* * *

And now I come to the recent emergence of Ebola and Zika.

I want you to continue to hold onto that image of an immune system orchestra without a conductor.

It may terrify you to know that I worked with the Ebola virus from 1992 to 1994 at Fort Detrick in a biosafety level 4 lab, the highest level of containment. The Zaire strain was being studied at the US Army Medical Research Institute of Infectious Diseases (USAMRIID), and it was my job to teach it how to infect human monocytes/macrophages without killing them. Because if you can't grow the virus, you can't study it.

It may surprise you to know that Ebola was never observed by Western medicine until 1976. Yes, that's right, hundreds of years of African exploration and development, and we never saw Ebola. But it must have been there, right? Yes, I'm sure it was, either in the bats or certain primates, but it never made that great leap from animals and into humans until after that time.

Again, I go to the CDC's own website for a brief history of Ebola:

> Ebola virus disease (EVD), one of the deadliest viral diseases, was discovered in 1976 when two consecutive outbreaks of fatal hemorrhagic fever occurred in different parts of Central Africa. The first outbreaks occurred in the Democratic Republic of Congo (formerly Zaire) in a village near the Ebola River, which gave the virus its name. The second outbreak occurred in what is now South Sudan, approximately 500 miles (850 km) away.
>
> Initially, public health officials assumed these outbreaks were a single event associated with an infected person who travelled between the two

locations. However, scientists later discovered that the two outbreaks were caused by two genetically distinct viruses: *Zaire ebolavirus* and *Sudan ebolavirus* . . .

Viral and epidemiological data suggest that Ebola virus existed long before these recorded outbreaks occurred. Factors like population growth, encroachment into forested areas, and direct interaction with wildlife (such as bush-meat consumption) may have contributed to the spread of the Ebola virus . . .

African fruit bats are likely involved in the spread of Ebola virus and may even be the source animal (reservoir host).[11]

Are we really to believe that in thousands of years of hunting, that Africans did not contract Ebola? It really seems comical to even suggest such a scenario. I believe these pathogens have likely been living in Africans for thousands of years until we did something to disturb the immune system balance of the people of that continent. The recent emergence of pathogenic Zika in Brazil and Columbia is also supported by that hypothesis.

What you do with a vaccination is you temporarily cripple a part of the immune system, as resources are diverted from protecting against other viruses to target the virus from the vaccine.

With multiple vaccinations, you cripple several parts of the immune system at the same time and do nothing to restore the balance of the system.

We don't know what diseases we are spreading by rendering compromised immune systems susceptible. It makes me angry because some of the best people in the world, like Christian missionaries and medical aid workers, are going to these countries and creating the conditions for terrible outbreaks. They are sending our very best people to unwittingly do the very worst things for the health of humanity.

Let's return to the CDC's own website for a listing of outbreaks and numbers of deaths to see if you can't discern a troubling pattern.

The Sudan outbreak of 1976 had 284 reported cases with 151 fatalities. The CDC website reports, "The outbreak is believed to have started with workers in a cotton factory where 37% of workers in the cloth room were infected."[12] Are we really to believe that workers at a cloth factory, of all the places in Africa, were the most likely to have been out in the bush hunting for monkeys and contracted this virus? Or is it more likely that just prior to coming down with Ebola, there was a workplace vaccination campaign? The other outbreak that year in Zaire had 318 reported cases with 280 deaths, although not much information is provided about the circumstances of that first appearance.

In 1977, there was only one person who contracted Ebola in Zaire and died.

We jump next to 1979, when there was another outbreak in the same area of the Sudan as the 1976 outbreak. The numbers, though, were much smaller this time. There were thirty-four reported cases with only twenty-two fatalities.

A full ten years pass until we see Ebola again, this time in 1989 and in two locations, one in the Philippines and one in the United States.

Both were . . . wait for it . . . facilities that housed monkeys.

Of the Philippine situation, the CDC wrote, "High mortality among Cynomolgus macaques was discovered in a primate facility responsible for exporting animals to the United States. Three workers in the animal facility developed antibodies, but never experienced symptoms of Ebola Virus Disease . . . Ebola-Reston virus was introduced into quarantine facilities in Virginia and Pennsylvania by monkeys imported from the Philippines."[13]

In 1990, this same Ebola-Reston virus was introduced to other quarantine facilities in Virginia and Texas by monkeys imported from the Philippines.

In 1992, Ebola-Reston virus was introduced into quarantine facilities in Siena, Italy, by monkeys from the Philippines.

In 1994, there was high mortality reported among chimpanzees in a forest in the Ivory Coast, and a scientist became ill after conducting an autopsy on a wild chimpanzee but later recovered. During that same year, an outbreak occurred in several gold-mining villages in the rainforest around Makakou, Gabon. There were fifty-two reported cases and thirty-one deaths.

In 1995, there was an outbreak associated with a charcoal maker in the forested area around Kikwit, Democratic Republic of the Congo (formerly Zaire). There were 315 reported cases and 250 deaths.

In 1996, there were several small outbreaks. There was one reported case from Russia, where a Russian laboratory worker was infected while working on an experimental treatment for Ebola. There were two outbreaks in monkey facilities in the Philippines and in Texas, with monkeys imported from the Philippines.

All was calm for the next four years until 2000, when in Uganda there was an outbreak of 425 reported cases and 224 deaths.

In 2001, there was an outbreak in the Republic of the Congo with fifty-nine cases and forty-three deaths and one in Gabon with sixty-five cases and fifty-three deaths.

In 2002, there was another outbreak in the Republic of the Congo with 143 reported cases and 128 deaths.

In 2003, an outbreak in the Republic of the Congo had thirty-five reported cases and twenty-nine deaths.

In 2004, a Russian laboratory worker who was working on an Ebola vaccine was accidentally injected with the virus and died. In the Sudan, there were seventeen reported cases and seven deaths.

In 2005, a small outbreak occurred starting with two hunters in the Republic of the Congo. Twelve people were affected and there were ten deaths.

No cases were reported in 2006, but in 2007 a new strain appeared in Uganda, which was significantly less lethal. There were 131 cases, but only forty-two deaths. In the Republic of the Congo there were 264 cases and 187 deaths.

In 2008, there was another small outbreak in the Republic of the Congo, with thirty-two cases and fifteen deaths. In the Philippines, the Ebola virus jumped to pigs and infected six workers at a pig farm, but they did not develop symptoms.

Nothing was reported for three more years until 2011, when a single person was infected with Ebola in Uganda and died.

In 2012, there was an outbreak in Uganda that affected six people and caused three deaths. In the Republic of the Congo, there was an outbreak with thirty-six cases and thirteen deaths.

Nothing happened in 2013, but 2014 saw the largest number of Ebola cases ever. In the Republic of the Congo, there were sixty-nine cases and forty-nine deaths. In the West African countries of Guinea, Liberia, and Sierra Leone, there were 28,610 and 11,308. In Italy, there was one case of an Italian healthcare worker who'd volunteered during the epidemic, but he survived. In Mali, an infected traveler brought the disease, resulting in eight cases and six deaths. In Nigeria, an infected traveler was responsible for twenty cases and eight deaths. In Senegal, an infected traveler was responsible for one case, but no deaths. In Spain, a healthcare worker became infected while treating a patient evacuated from Sierra Leone, but he recovered. In the United States, there were four confirmed cases, with two being nurses who were treating an Ebola patient on American soil. One person died in the United States outbreak.

Next, we jump forward three years to 2017, when there were eight cases and four deaths in the Republic of the Congo.

In 2018, there was another outbreak in the Republic of the Congo with fifty-four cases and thirty-three deaths. The World Health Organization declared the outbreak over on July 24, 2018. This was the ninth recorded outbreak of Ebola in the Republic of the Congo.

In just a little over two pages, I've provided the approximately twenty known outbreaks of Ebola and a brief description of the circumstances of each outbreak. Now, let's really try to consider what's happening.

* * *

Let's leave aside for a moment the question of why this disease suddenly appeared in 1976. Instead let's focus on the ten instances of Ebola appearing in animal or scientific facilities.

In 1989, there was an outbreak of Ebola at a primate facility in the Philippines.

In 1989, there was an outbreak of Ebola at primate facilities in Virginia and Pennsylvania.

In 1990, there was an Ebola outbreak at monkey quarantine facilities in Virginia and Texas.

In 1996, a Russian lab worker was infected with Ebola while working on an experimental treatment.

In 1996, there was an Ebola outbreak in monkey facilities in the Philippines.

In 1996, there was an outbreak of Ebola at monkey facilities in Texas with monkeys imported from the Philippines.

In 2004, there was another laboratory worker in Russia who was accidentally injected with the virus and died.

In 2008, the Ebola virus somehow made a jump into pigs and from there infected six workers, but none of them developed symptoms.

Is it just my imagination, or are some of the most likely places on Earth to contract Ebola a scientific lab or hanging around with monkeys in cages, presumably in unhealthy conditions? Not monkeys in the wild.

Let's move onto the West African outbreak.

In a little bit of research, we came across an interesting account written by Hong Kong-based journalist and former editor of The Japan Times in Tokyo, science writer Yoichi Shimatsu.

> The mystery at the heart of the Ebola outbreak is how the 1995 Zaire (ZEBOV) strain, which originated in Central Africa some 4,000 km to the east in Congolese (Zairean) provinces of Central Africa, managed to suddenly resurface now a decade later in Guinea, West Africa. Since no evidence of Ebola

infections in transit has been detected at airports, ports or highways, the initial infections must have come from either one of either two alternative routes . . .

The reason for suspecting a vaccine campaign rather than an individual carrier is due to the fact that the Ebola contagion did not start at a single geographic center and then spread outward along the roads. Instead, simultaneous outbreaks of multiple cases occurred in widely separated parts of rural Guinea, indicating a highly organized effort to infect residents in different locations in the same time frame.

The Ebola outbreak in Early March coincided with three separate vaccination campaigns countrywide: a cholera oral vaccine effort by Medicins Sans Frontieres under the WHO; and UNICEF-funded prevention programs against meningitis and polio.[14]

Let's just say that these claims might not be able to be substantiated.

However, it seems clear to me that there were three separate vaccination campaigns PRIOR to the West African Ebola outbreak. You simply have to consult news accounts to learn about all of these vaccination campaigns.

Maybe that was a coincidence. I don't believe in coincidences.

* * *

I've talked about how one of the major problems with multiple vaccinations is the so-called "war and Peace of the viruses" in which each virus preoccupies a different part of the immune system, leading to a compromised immune system. I give you this publication from the CDC titled "Emergence of Vaccine-Derived Polioviruses during Ebola Virus Disease Outbreak, Guinea, 2014–2015." Did you catch that term "Vaccine-Derived Polioviruses"? And it happened during an Ebola outbreak. Not among those who came down with the disease, but simply in the community. The abstract reads:

During the 2014–2015 outbreak of Ebola virus disease in Guinea, 13 type 2 circulating vaccine-derived polioviruses (cVDPVs) were isolated from 6 polio patients and 7 healthy contact. To clarify the genetic properties of cVDPVs and their emergence, we combined epidemiologic and virologic data for polio cases in Guinea. Deviation of public health resources to the Ebola outbreak disrupted polio vaccination programs and surveillance activities, which fueled the spread of neurovirulent VDPVs in an area of low vaccination coverage and immunity.[15]

There are times when my dear colleagues in science make me want to bang my head against a wall. Let's take that final sentence from the abstract: "Deviation of public health resources to the Ebola outbreak disrupted polio vaccination programs and surveillance activities, which fueled the spread of neurovirulent VDPVs in an area of low vaccination coverage and immunity."

Let's translate that mumbo-jumbo into something understandable. Then you'll realize how that sentence makes no sense.

It should read something like this: "Because of the Ebola outbreak we couldn't give our polio vaccines, and that promoted the spread of polio viruses from our vaccines." Really, I encourage you to come up with any other translation.

Here's what it really means to me. Our vaccines are generating new viruses, and in areas with low immune function, think poor and impoverished (or they're getting so many vaccinations that their immune system is going haywire trying to keep up), we're making sure these people will get some type of serious viral diseases like HIV, Ebola, or Zika.

* * *

Probably no name is more associated with the West African outbreak of Ebola than Dr. Kent Brantley, the US doctor who was working in Liberia during the outbreak, contracted Ebola, and was flown back to the United States for treatment and recovered. I have a great deal of respect for Dr. Brantley, a strong Christian who believes his ministry is healing people, and I found the account of his ordeal in the book *Called for Life* to mirror many of my own beliefs.

In the prologue to the book, Brantley wrote, "For the thirty-eight years since Ebola Virus Disease had been identified, every outbreak had been limited to small rural communities. This time, however, was different. This time, Ebola had found the perfect storm of factors, quickly spreading through three countries and into major urban centers."[16] Farther on, his account of the number of outbreaks is in accord with what I understand: "There had been fewer than twenty Ebola episodes since the virus was first identified in 1976. . . . The most deaths from an Ebola outbreak had been 280 in Zaire in 1976."[17]

As to what caused the outbreak to be so severe in West Africa, Brantley writes about the main city of Monrovia in Liberia: "In the city, there was limited access to clean water, so many people could not practice proper hygiene."[18] I would also add that West Africa has been through years of instability with various wars, leaving the people in a nutritionally depleted state and vulnerable to pathogens that might be activated.

When Dr. Brantley was diagnosed with Ebola, he contacted somebody he had recently met, Dr. Randy Schoepp, chief of diagnostics with USAMRIID. You'll forgive me, but after what I've learned over the years about what military infectious medical researchers do from my experiences at Fort Detrick, I'm a little wary of any individuals in that kind of a position.

Schoepp went through the options with Brantley, eventually settling on a cocktail of antibodies that were believed to target the Ebola virus. This is what Brantley wrote, and I'm sure you'll understand immediately what concerns me so much:

> Because the drug had not been administered to a human, there was no way of knowing whether it was safe. The serum was derived from a mouse antibody and grown in tobacco plants. It is a type of monoclonal antibody. I knew that monoclonal antibodies were engineered in laboratories to imitate antibodies naturally produced as part of the body's immune system. That type of treatment had been employed to treat other conditions, and as far as I knew they were generally considered to be safe.[19]

Yes, we've got our lovely little mouse again, and we don't know what viruses are sleeping in that little rodent, ready to wake up and wreak havoc on the human immune system. And all of that faith in biological products that are "engineered in laboratories" and "generally considered to be safe" is just a little out of my comfort zone. Still, if I'm looking at likely death from Ebola, I'll take the option that gives me the best chance for survival.

But we need to know how the whole disaster got started. We owe that to humanity.

I think the heart of this darkness lies not in the African jungles, but in our own research labs. We play God without telling the truth about our failures.

* * *

But there is little appetite for doing good science about vaccines. Instead, the scientific media simply wants to shout down anybody who asks questions. What happened to persuasion as the appropriate method of civil discourse in our society? When did we suddenly become authoritarian? When did dialogue and listening to one another become a sign of weakness?

Consider this recent article from *Scientific American* on March 21, 2019. The title is "Opting Out of Vaccines Should Opt You Out of American Society."[20] The subheading reads, "People who are able to take vaccines but refuse to do so are the moral equivalent of drunk drivers." Wow, that's an amazing moral comparison. I guess all my years of scientific study and research have made me drunk with curiosity!

Here's a sad sample of what passes for discussion in what used to be one of the premier magazines for discussion of scientific progress:

> There is no moral difference between a drunk driver and a willfully unvaccinated person. Both are selfishly, recklessly and knowingly putting the lives of everyone they encounter at risk. Their behavior endangers the health, safety and livelihood of the innocent bystanders who happen to have the misfortune of being in their path.
>
> The reasons why are simple and straightforward. Vaccines aren't perfect (e.g., they can wear off over time) and not everyone can be vaccinated. There is one and only one reason to skip a vaccine: being immunocompromised. Some individuals, because of genetic deficiencies or diseases like cancer, cannot receive vaccines. Other people are too young. Vaccines such as MMR (measles, mumps, rubella) cannot be administered before 12 months of age.[21]

As a person trained in the skill of scientific argumentation, the two quoted paragraphs are almost painful to read. The first paragraph has nothing substantial in it, just fearmongering. Let's move onto the second paragraph.

Vaccines aren't perfect. Big understatement.

But what do they discuss as the only problem? They wear off over time. That's kind of like saying the only thing wrong with chocolate is that when you finish eating it, there's nothing left. I'd like a little more discussion.

Color me a curious scientist.

Just to be brief about the issue, I think one of the major problems with vaccines is that they're grown in animal tissues and we don't know what viruses and pathogens are coming back in the needle. A recent inquiry in December 2018 by the Italian lab, Corvela, on the GlaxoSmithKline vaccine Priorix Terta highlights troubling problems that our technology can now uncover but that few seem to have the courage to investigate. Translated from the Italian, the report finds:

> We have continued the investigation, both chemical and biological, on the Priox Tetra, quadrivalent against measles, rubella, mumps, and varicella. We

have found . . . proteobacteria and nematoda worms, 10 other viruses through ssRNA, Microviridae (bacterial or phage viruses) and numerous retroviruses including endogenous human and avian retroviruses, avian viruses, human immunodeficiency and immunodeficiency virus of monkeys (fragments that if inserted into the database detect fragments of HIV and SIV), murine virus, horse infectious anemia virus, lymphoproliferative disease virus, Rous sarcoma virus, alphaendornavirus, hepatitis B virus, and yeast virus.[22]

The technology exists to answer the question of how many different viruses are contaminating our typical vaccines. Look at the problems with this single vaccine. There are indications of viruses from humans, birds, monkeys, mice, and horses, all animals that we use in one way or another in vaccine production. And what about the worms, yeast, and other microorganisms that naturally occur in animal tissue?

If you eliminate the animal tissue, that leaves aborted human fetal tissue, and I think there are significant moral and scientific issues with what happens on a genetic level when you inject human tissue into the bloodstream. Then you get to the issue of chemicals in the vaccines, like mercury, aluminum, formaldehyde, polysorbate 80, and a host of others, and it begins to look like a witch's brew that would only be given to children in some demented fairy tale.

Personally, I feel like I'm one of the few sober drivers on the road when it comes to the question of vaccines and human health.

CHAPTER TEN

My Coauthor Gets Banned from Australia

Sometimes you need a little levity, or you'll just go crazy.

I'm happy that my coauthor, Kent, even though he lives a difficult life with a severely autistic daughter, keeps me laughing at the absurdity of much of what is going on in science today, and specifically in the scientific press.

If I didn't know any better, I'd claim he was an undiscovered member of the satirical English group, Monty Python, since, as in their movie *Life of Brian*, a parody of the Jesus Christ story, I imagine him whistling the song "Always Look on the Bright Side of Life," even as he's being crucified in the media.

Do you think I'm kidding?

Let me tell you a story.

* * *

Science is supposed to work in a certain predictable way.

A hypothesis is made, then it's investigated in an objective manner. If the data can be replicated, it becomes an accepted fact. If not, then the hypothesis is disproven.

This isn't always the case. Controversies and scientists with diametrically different findings have erupted in many fields, such as the health effects of smoking, leaded gasoline, pesticides, and the possible negative reactions of certain pharmaceutical drugs.

One might say when large amounts of money are involved, the truth is in danger.

In 1998, Dr. Andrew Wakefield and eleven other authors published a blockbuster article in *The Lancet*, a well-known medical journal, alleging a possible association between the MMR (measles-mumps-rubella) vaccine, the development of autism, and a specific type of gastrointestinal problem. In other words, the vaccine was doing something that affected the gut, which in turn was affecting the brain. Despite what you may think you know about Dr. Wakefield, if you've ever used the expression "gut-brain connection," meaning there's a link between digestive issues and mental health, you have Dr. Wakefield to thank for that contribution to science. As Dr. Wakefield explained to Kent:

> The paper described clinical findings in twelve children with an autistic spectrum disorder (ASD) occurring in association with a mild-to-moderate inflammation of the lymph glands in the intestinal lining (lymphoid nodular hyperplasia), predominantly in the last part of the small intestine (terminal ileum). Contemporaneously, parents of nine children associated onset of symptoms with measles, mumps, rubella (MMR) vaccine exposure, eight of whom were reported on the original paper.[1]

On January 28, 2010, after years of heated controversy, the General Medical Council of Great Britain found Dr. Wakefield guilty of unethical behavior in his research. Despite the accusations of fraud made in the media, the claim against Dr. Wakefield in the complaint was for "callous disregard" of children for submitting them to the standard tests for gastrointestinal problems. When Dr. Wakefield was struck off the register of physicians in May of that year, this is what the *Daily Mail* reported about the case:

> The panel said he behaved unethically and showed "callous disregard" for any distress of pain the children might suffer.[2]
>
> The children underwent typical gastro-intestinal tests, such as colonoscopies, lumbar punctures, and barium meals, in addition to urine tests. All of this is standard practice, but to the experts of the General Medical Council it was akin to torture.
>
> The panel ruled that many of the children should never have been included in the research. It also found that Dr. Wakefield and his colleagues had not been granted ethical approval to use the children in their research.

Professor Terrence Stephenson, President of the Royal College of Pediatrics and Child Health, said Dr. Wakefield's research had caused "untold damage."[3]

Does that make any sense to you?

If a physician thinks there may be problems in your gastrointestinal system, you'll get a colonoscopy. Just ask any man over the age of fifty. That's when you're supposed to get your first one. Okay, and a urine test is among those things that are supposed to cause "pain and distress"?

None of the tests given to the children were outside the standard of practice for suspected gastrointestinal problems.

And as a later appeal found, ethical approval had been given for those children. They were being treated for gastrointestinal problems, for God's sake! Does one refuse to look at data because they were obtained in the course of treatment? That's called standard medical practice.

Dr. Wakefield was not stripped of his license because of fraud. If you believe the reporting on this issue, he was stripped of his license because he ordered some unnecessary tests.

And in addition, even though it was the longest and most costly hearing in the more than 150-year history of the General Medical Council, none of the parents could testify in support of Dr. Wakefield.

In fact, the Council did not allow any of the parents to speak.

How can you be a supervillain if there are no victims?

* * *

One might ask what happened in the twelve-year interval between the publication of Wakefield's article in *The Lancet* and being struck off the register of physicians in 2010. Surely there were some high-level investigations by independent investigators into the claim of an autism risk from the MMR vaccine?

The answer would be, yes, there was a high-level investigation by the Centers for Disease Control (CDC), published in 2004, and known as the DeStefano paper, after the lead author, Frank DeStefano. Surprisingly, the study did show an increased risk, but below the level necessary to ring any causation alarm bells.

But in November 2013, a remarkable thing took place. Dr. Brian Hooker, an autism parent and longtime critic of the CDC for what he believed to be a lack of transparency on the issue, was contacted by Dr. William Thompson, one of the authors of the DeStefano MMR study.

Apparently, Thompson had been carrying a great secret for many years.

Thompson claimed that their research uncovered an association between earlier administration of the MMR vaccine and rates of autism, particularly among African American males. Hooker didn't want this potential whistle-blower to vanish into the night, so Hooker contacted an attorney to see how he might legally record these conversations.

Hooker was told that the laws of Washington State allowed one party in a telephone conversation to record what was being said, so Hooker traveled several times from his home in the far north of California to a location in Washington State. Once Hooker had several taped conversations, he contacted Dr. Andrew Wakefield, who had devoted himself to fighting for children with autism and clearing his name.

This was bombshell material.

Eventually, they would work with Del Bigtree, an Emmy-winning Hollywood television producer who had worked on *The Dr. Phil Show* as well as the medical talk show *The Doctors*. This group would go on to produce a documentary about this case titled *VAXXED: From Cover-Up to Conspiracy*. The relationship between Hooker and Thompson grew so close that Hooker was able to convince Thompson to apply for federal whistle-blower protection and release a statement through his attorney to the press about his claims.

Thompson had been deeply troubled over the years by what he had been forced to conceal and lamented at times to Hooker that now whenever he saw an autistic child, he felt guilty and responsible.

The trove of documents Thompson had retained was turned over to Congressman William Posey (R) Florida, who offered to make them available to any interested journalist. My coauthor applied for and was granted access to those documents, eventually writing a book about the Thompson case and other aspects of the vaccine issue titled *INOCULATED: How Science Lost Its Soul in Autism*. It's a fine book, and I suggest you read it.

To me it's more terrifying than any Stephen King novel.

Thompson would eventually write a long and detailed confession for Congressman Posey. Here is the concluding section, in Dr. Thompson's own words, about how the CDC actively covered up evidence of a linkage between the MMR vaccine and autism in the years between 2001 and 2004:

> I believe we intentionally withheld controversial findings from the final draft
> of the De Stefano et al (2004) Pediatrics paper. We failed to follow the final
> approved study protocol and we ran detailed in depth RACE analyses from

October 2001 through August 2002 attempting to understand why we were finding large vaccine effects for blacks. The fact that we found a strong statistically significant finding among black males does not mean there was a true association between the MMR vaccine and autism-like features in this subpopulation.

This result would probably have led to designing additional better studies if we had been willing to report the findings in the study and the manuscript at the time we found them. The significant effect of early vaccination with the MMR might also have been a proxy for the receipt of thimerosal vaccine early in life but we didn't have the appropriate data to be able to code the level of thimerosal exposure from the MADDSP school records.

In addition to significant effects for black males, we also found significant effects for "isolated autism cases" and for the threshold of 24 months of age. If we had reported the 24 month effects, our justification for ignoring the 36 month significant effects would not have been supported. In the discussion section of the final published manuscript, we took the position that service seeking was the reason we found a statistically significant effect at 36 months. This was a post-hoc hypothesis regarding the findings after we confirmed one of our primary hypotheses.[4]

Leading scientists at the CDC charged with ensuring the safety of vaccines knew beyond a shadow of doubt in 2004 that earlier administration of the MMR vaccine was leading to increased rates of autism in African American males and to a lesser extent among the population at large. It is difficult to imagine a greater crime against the health of the public.

It seems to me that Thompson details five specific lies in his confession.

The first lie is the failure to report a significant effect for African American males who received an MMR vaccine at twelve months instead of thirty-six months.

The second lie is the failure to report significant effects for "isolated autism cases," meaning the sudden and unexpected development of autism in children with no previous behavioral or health problems.

The third lie is the failure to report significant effects for those children whom received the MMR vaccine at twenty-four months rather than twelve or thirty-six months.

The fourth lie is that even with the removal of a significant number of African-American boys with autism from the study because they did not have a Georgia birth certificate, a statistically significant effect remained, but they claimed this was likely because of a result of "service-seeking" and a mythical regulation that a condition of such services was the earlier administration of the MMR shot.

The fifth lie was that by omitting all of this information, it lowered concern about damage caused to children of all races from the MMR vaccine. All of this took place within the 2001–2004 time frame.

Would Wakefield have been the subject of an inquiry by the General Medical Council in 2010 if the CDC scientists had honestly reported their findings in 2004?

Probably not.

Would Frank and I have been attacked so viciously from 2010 to 2012 for my concern about the role vaccines and retroviruses might be playing in the development of autism?

Probably not.

Dr. Thompson continues to work at the CDC and has been waiting to be called to testify in front of Congress. Because of the stranglehold the pharmaceutical companies have on our elected representatives, I doubt his story will ever be shared by the mainstream media.

* * *

There are a few individuals in Congress who have tried to raise questions about vaccines. Even if these members of Congress hold great power in other spheres, it seems that when they start talking about vaccines, they get no coverage from the media. Former Congressman Dan Burton did great work, as well as Congressman and physician Dave Weldon. Current Congressman Bill Posey continues to try and get this story out to the public.

On July 29, 2015, Posey took to the floor of the House of Representatives to speak about the active concealment of the CDC regarding this information. He was only given five minutes to speak, but he packed in a great deal.

> Mr. Speaker, I rise today on matters of research and scientific integrity.
>
> To begin with, I am absolutely, resolutely pro-vaccine. Advancements in medical immunization have saved countless lives and greatly benefited public health.
>
> That being said, it is troubling to me that, in a recent Senate hearing on childhood vaccinations, it was never mentioned that our government has paid out over three billion dollars through the National Vaccine Injury Compensation Program for children who have been injured by vaccinations.

Regardless of the subject matter, parents making decisions about their children's health deserve to have the best information available to them. They should be able to count on Federal agencies to tell them the truth.

For these reasons, I bring the following matter to the House floor. In August 2014, Dr. William Thompson, a senior scientist at the Centers for Disease Control and Prevention, worked with a whistleblower attorney to provide my office with documents related to a 2004 CDC study that examined the possibility of a relationship between the mumps, measles, and rubella vaccine and autism.

In a statement released in August of 2014, Dr. Thompson stated: "I regret that my coauthors and I omitted statistically significant information in our 2004 article published in the Journal of Pediatrics."

Mr. Speaker, also quoting Dr. Thompson:

My primary job duties while working in the immunization safety branch from 2000 to 2006 were to lead or co-lead three major vaccine safety studies. The MADDSP MMR-Autism Cases Control Study was being carried out in response to the Wakefield Lancet *study that suggested an association between the MMR vaccine and an autism-like health outcome.*

There were several major concerns among scientists and consumer advocates outside the CDC in the fall of 2000 regarding the execution of the Verstraeten study.

One of the important goals that was determined upfront in the spring of 2001 before any of these studies started was to have all three protocols vetted outside the CDC prior to the start of the analyses so that consumer advocates could not claim that we were presenting analyses that suited our own goals and biases.

We hypothesized that if we found statistically significant effects at either 18- or 36-month thresholds, we would conclude that vaccinating children early with MMR vaccine could lead to autism-like characteristics or features.

We all met and finalized the study protocol and analysis plan. The goal was not to deviate from the analysis plan to avoid the debacle that occurred with the Verstraeten Thimerosal study published in Pediatrics *in 2003.*

At the September 5 meeting, we discussed in detail how to code race for both the sample and the birth certificate sample. At the bottom of table 7, it also shows that for the nonbirth certificate sample, the adjusted race effect statistical significance was huge.

All the authors and I met and decided sometime between August and September 2002 not to report any race effects for the paper. Sometime soon after the meeting, where we decided to exclude reporting any race

effects, the coauthors scheduled a meeting to destroy documents related to the study.

The remaining four co-authors all met and brought a big garbage can into the meeting room and reviewed and went through all the hard copy documents that we had thought we should discard and put them into a huge garbage can.

However, because I assumed it was illegal and would violate both FOIA and DOJ requests, I kept hard copies of all documents in my office, and I retained all associated computer files.

I believe we intentionally withheld controversial findings from the final draft of the Pediatrics *paper.*

Mr. Speaker, I believe it is our duty to ensure that the documents Dr. Thompson provided are not ignored; therefore, I will provide them to Members of Congress and the House Committees upon request.

Considering the nature of the whistleblower's documents, as well as the involvement of the CDC, a hearing and a thorough investigation is warranted.

I ask, Mr. Speaker, I beg, I implore my colleagues on the Committee on Appropriations to please, take such action.[5]

Congressman Posey might as well have been speaking in that proverbial forest in which no person hears the fall of a tree, for all the media coverage generated by his remarks.

Posey eventually made his documents available to any member of the press who requested them. As far as I know, my coauthor, Kent, is the only one who ever requested them. Maybe Sharyl Attkisson asked for them as well, I don't know.

Let's review what Thompson had to say.

He had management responsibilities for three major vaccine safety studies for the Centers for Disease Control and Prevention. I think that would qualify him as an expert to discuss the reliability of the government's science.

Thompson relates that the study design was created and approved with the input of scientists and consumer advocates. Got it? Both scientists and the parents of vaccine-injured family agreed on the study design.

When they looked at earlier and later administrations of the MMR vaccine, they found a statistically significant race effect, which was in his words was "huge."

These five scientists at the CDC decided not to report the race effects from earlier administration of the MMR vaccine. And let's not forget, they were only looking at children who got an earlier dose of the MMR vaccine and those who got a later dose. Anybody with just a passing familiarity with science knows that in order to have a valid study you need to have a control group.

What does that mean?

It means you need to have a group of children who didn't get the MMR shot. The study they did is equivalent to looking for lung cancer rates among those who smoked one pack of cigarettes a day versus those who smoked two packs of cigarettes a day.

And just because I happen to believe that scientists should be thorough in their investigations, how about a study looking at a group of fully vaccinated kids and another group who did not get any vaccines? If you believe the CDC, the more educated and affluent you are, and the more you research vaccines, the more likely you are to alter the schedule.

As if it weren't bad enough that these five CDC scientists decided to cover up this information, they even made a Saturday afternoon date to destroy information.

How could this possibly be any worse? Can there be any bigger news story in the world than children being harmed when they walk into their pediatrician's office?

When the government investigated vaccines and thimerosal in the 2000–2003 time period, they found an increase in neurological problems, as the Verstraeten study showed. But of course, they washed that data to make the signal go away, triggering what they euphemistically called a "controversy." It weren't a "controversy," it was a crime.

When the government investigated the MMR vaccine and autism in the 2001–2004 time period, they found an effect, but they decided to cover it up, even going so far as to come into the CDC on a Saturday and try to throw away all the incriminating evidence into a large garbage can. We only know about this because of Dr. William Thompson's revelation from 2014, more than ten years after these incidents took place.

In 2007, the government's own medical expert in the Vaccine Court, Dr. Andrew Zimmerman, told the government that vaccines were causing autism in some children. Less than forty-eight hours after he provided this information, the government kicked him off the case. We only know this because of an affidavit Zimmerman signed in 2018.

Let's just say it.

Every time the government investigates vaccines, they find problems and they make the decision to cover up the information.

* * *

And now we come to the Australia part of the story.

After Dr. Brian Hooker contacted Dr. Andrew Wakefield about the allegations of Dr. William Thompson, they made a documentary film about the case called *VAXXED: From Cover-Up to Catastrophe,* which featured several of the legally recorded conversations between Dr. Thompson and Dr. Brian Hooker. One of the producers was an Englishwoman, Polly Tommey, the mother of an autistic child and cofounder of *Autism File* magazine. Kent had written several articles for *Autism File,* so he had an acquaintance with Polly.

The film was controversial to people who never actually sat in a theater and watched it, most famously being accepted into the prestigious Tribeca Film Festival, then getting disinvited when protests were made to festival founder Robert De Niro. So much for artistic freedom! It was not De Niro's finest hour. It was revealed at the time that De Niro had an eighteen-year-old son with autism, and his wife said he changed after a vaccination. Apparently, De Niro had been away filming and didn't feel he could comment with any accuracy on his wife's claim. Way to support your wife! Perhaps it does not come as a surprise that Mr. De Niro and his wife broke up a few years later?

The *VAXXED* crew took their documentary to Australia, only to find themselves harassed by the press and government, who hadn't expected an antivaccination film to land in their country. The media and government harassment were game changers for the tour, since before the attacks they hadn't been selling many tickets. After the media firestorm, they sold out all the venues.

The Australian government wasn't done with the *VAXXED* crew. When they went to leave the country, they were detained by immigration while their materials were searched and photographed. Since content was obviously not what they were looking for, perhaps it was links or information on their own indigenous anti-vaxxers. At the airport, as they were finishing up with immigration, one of the agents told Polly Tommey, "If we'd known you were coming, we wouldn't have let you in the country."

When he heard about this, Kent quickly dialed Polly, reaching her in New Zealand, and conducted a quick interview for an article in *Bolen Report,* where he regularly writes. Polly told Kent, "We were in the papers every day. They're their own worst enemy. They promoted the fact we were in Australia and the talks the whole time. If they had shut up, no one would have even known we were there. They did this to themselves. And the more they kept saying *VAXXED* is dangerous and a pack of lies, of course, the more people wanted to see it. They were idiots to promote us like they did. We were nowhere near sold out before we arrived. But the minute we were in Australia and the press got the story, we sold out every venue. We were packed and had a waiting list of people wanting to see *VAXXED*."[6]

All of this gave Kent an idea.

I learned long ago that Kent and an idea are an interesting combination.

* * *

Even though Kent had no interest in traveling to Australia, he applied for a visa to travel there on a series of speaking engagements, which he dubbed the "Dangerous Science" tour. He had no speeches scheduled, venues, and hadn't even contacted a single person in Australia. He announced it with great fanfare in the pages of *Bolen Report*, even publishing an open letter to Australian Prime Minister Malcolm Turnbull detailing his planned visit and what he hoped to accomplish. This is what my trouble-making coauthor wrote:

> Dear Prime Minister Turnbull:
>
> I am very excited to visit your country in December of 2017 on my **"Dangerous Science"** tour in which I will be discussing my two books, *PLAGUE: One Scientist's Intrepid Search for the Truth About Retroviruses* and *INOCULATED: How Science Lost its Soul in Autism*. I will also be discussing my recent efforts to bring about a FIVE-YEAR MORATORIUM ON CHILDHOOD VACCINES in the United States through both a White House and FDA/CDC petition by way of the Administrative Practices Act.
>
> If you wish to obtain more information about this effort, you can go to tinyurl.com/vaccine moratorium.
>
> Since you are not an American citizen, you cannot sign either of the petitions. I wouldn't want you to get in trouble, but perhaps it might give you some ideas for similar legislation in Australia.
>
> I understand there is a great deal of confusion over the nomenclature among many in our community, with some wishing to be **"pro-safe vaccine"** or **"vaccine safety"**, but after much study in this field, I think I should simply be referred to as the **"World's #1 Anti-Vaxxer."** I understand there are others who may feel they deserve the title, and I am happy to share it, but let's simply use it for the sake of convenience.
>
> I think you would find my book, *INOCULATED: How Science Lost its Soul in Autism*, to be especially revealing as it is based on documents

provided to me by US Congressman William Posey, who in turn received these documents from whistle-blower, Dr. William Thompson, a vaccine safety scientist at our Centers for Disease Control and Prevention (CDC). Thompson is currently waiting to be called to testify in front of the House Oversight and Government Reform Committee in our Congress. He has been waiting for three years. I hope things move quicker in Australia, especially considering Thompson's testimony about how the MMR vaccine is devastating our African-American population. I imagine it is also doing the same to your aboriginal population.

My intention in coming to Australia is to make both *PLAGUE* and *INOCULATED* the two best-selling books in Australian history, as well as to create a movement for a similar Australian **FIVE-YEAR MORATORIUM ON CHILDHOOD VACCINES**.

If you wish to greet me at the airport when I land, I would be amenable to such an effort, or if you think our meeting should be at The Lodge, the primary residence of the Prime Minister, I would accept that as well. After all, I will be a guest in your country.

Please don't make too much of a fuss over my efforts, but if you would like to give me an award in recognition of my work to liberate Australia from the grip of the pharmaceutical industry, I like the sound of **The Order of Australia, Officer of the Order.** Maybe it's the imperialist in me, but it just sounds so regal.

I'm also enclosing two articles for your review, "**Australia Bans Autism Mom as an 'ENEMY OF THE STATE'**" (*Bolen Report*, August 10, 2017) and "Kent Heckenlively—World's #1 Anti-Vaxxer" (*Bolen Report*, August 8, 2017) so you will have greater familiarity with my work.

I look forward to visiting your beautiful country and meeting the brave citizens of Australia!

Your humble servant,
Kent Heckenlively

P.S.—I really LOVE your new automated visa application form. I should already be approved by the time you get this letter. Throw a shrimp on the barbie for me, mate![7]

Kent thought it was a funny piece that would be appreciated by his fellow members in the health freedom community but never expected it to become an international story plastered in papers across the globe.

* * *

The response was swift and hilarious.

In the *Sydney Morning Herald* on August 31, 2017, there was a scream-ing headline that pronounced, "'World's Number 1 Anti-Vaxxer' Kent Heckenlively Denied Entry to Australia." The article read:

> The self-proclaimed "world's number one anti-vaxxer" has been denied per-mission to visit Australia.
>
> Immigration Minister Peter Dutton said on Thursday Kent Heckenlively would not be able to tour Australia later this year as part of an international campaign calling for a pause in childhood vaccinations.
>
> "We're not going to allow him to come here," Mr. Dutton told Sydney radio station 2GB.
>
> "These people who are telling parents that their kids shouldn't be vac-cinated are dangerous. We have been very careful in having a look right through this particular case and it's clear to me that It's not in our national interest that he should come here."[8]

Wow, so you can be a socialist in America, committed to the overthrow of the capitalist system, get elected to the United States Congress, and dominate the Democratic Party. I'm sure that Alexandria Ocasio-Cortez would have no problem getting a visa to Australia. But let somebody ques-tion a pharmaceutical product made by a company that contributes massive amounts of money to media companies and politicians, and even the immi-gration minister of the country will get involved.

The story of Kent's banning from Australia was covered not just in Australian papers, but on the *BBC*, the English newspaper *The Guardian,* the *International Business Times, Buzzfeed* in America, and the *South China Post*. There are more than five hundred news articles you can find on Kent and Australia in Google news. Maybe that's a healthy discussion, but Kent has never received a single phone call from any of those national and international newspapers asking for his side of the story, let alone a comment.

The documents received by Kent from the Australian government were interesting, as it appeared the country had originally granted his application but then later rescinded the approval. In a letter he received on August 31, 2017, it read: "Dear Mr. Heckenlively, I wish to advise you that your sub-class 601 Electronic Travel Authority (ETA) visa granted on 13 August 2017 was cancelled on 31 August 2017 under section 128 of the Migration Act 1958 ('the Migration Act')."[9]

As I read the response, it seems Kent's visa was approved on August 13, then rescinded more than two weeks later.

It may seem a little dry to quote extensively from a bureaucratic document, but it's critical because these are the rules under which governments are deciding which conversations are allowed to take place. Please be aware that Kent has never been convicted of a crime, and everybody who meets him considers him kind and well spoken.

> On 14 August 2017 the Department received information from Mr. Heckenlively stating that he intended to tour Australia in December 2017 to promote his views on anti-vaccination and promote his anti-vaccination books.
>
> Open source information indicates that Mr. Heckenlively also plans to discuss during his tour of Australia how the pharmaceutical industry and government entities have made human lives more dangerous. He is also quoted as saying he regards himself as the "World's Number One Anti-Vaxxer."
>
> Open source information from the scientific and medical fraternities in Australia states that vaccinations are essential to ensure the health of the Australian community is protected, and that the Australian community's health would be seriously threatened without the immunization program conducted in Australia. The Australian Government's Immunize Australia Program funds free vaccination programs, administers the Childhood Immunization Register and provides information to the Australian public and health professionals.
>
> The Commonwealth Department of Health's website provides that immunization is the safest and most effective way of protecting children against disease and with sufficient numbers of people immunized in the Australian community helps prevent the spread of these diseases, or is eliminated altogether.
>
> Based on this information, I am satisfied that Mr. Heckenlively's intentions to promote his views on anti-vaccination would or might be a risk to the good order of the Australian community.[10]

My dear coauthor, Kent, is a threat to the "good order" of a strong and thriving democracy? He's listened to me for countless hours and written books with hundreds of footnotes to support his claims. And he gets ridiculed by his family for driving the speed limit, uses the crosswalk when crossing the street, and flosses every night. The book he wrote about corruption in the vaccine program of the CDC was based on documents he received from the United States Congress. He is advocating for good science being applied to looking at the effects of vaccination, an investigation that is not currently being done.

Now, the Australian authorities might say Kent acted a little disrespectfully in his letter. We know the leader of any country should be addressed as "Your Most High and Exalted Excellency and Ruler of All."

That's just proper diplomacy.

And he did call for a naval blockade of the country until democracy was restored, as well as the provision of nonlethal aid to the Australian anti-vaxx rebels. But those last two calls were issued only AFTER they banned him from Australia. And before issuing these calls, Kent did send them a copy of his book *INOCULATED: How Science Lost its Soul in Autism* and asked them to point out any errors of fact that rendered him unworthy of traveling to Australia.

And he did threaten in an article to secretly travel to Australia under a false name and wearing a disguise, which in truth was a pair of phony black-rimmed glasses with a big nose and bushy eyebrows. He even published a photo of himself in the planned disguise in front of a map of Australia but instructed all his readers not to let the Australian authorities know.

The government did not initiate any further contact with Kent.

But the final indignity was yet to come.

* * *

Kent was in Florida, where his wife was going to be running a half-marathon with another member of her family, when he got a call from his brother in California. Kent's brother was laughing so hard over the phone that Kent could barely make out what he was saying.

"R. Kelly got banned from Australia," Kent could finally understand what his brother telling him. "It's in the *Sacramento Bee*."

Kent had some vague recollection that R. Kelly was a musician who'd been accused of inappropriate behavior with young girls. "Okay, so what does that have to do with me?"

"They mention you in the article about R. Kelly!" he was eventually able to say through the laughter.

Kent quickly went to the Internet and typed in his name along with R. Kelly and found an article from the *Hollywood Reporter:*

Kelly's career has been stifled since a #MuteRKelly campaign gained momentum last year to protest his alleged sexual abuse of women and girls, which
Kelly denies. Lifetime's documentary series *Surviving R. Kelly* last month
drew even more attention to the allegations, and his record label has report-
edly dropped him.

Australia has denied entry to other foreigners on character grounds,
among them troubled R&B singer Chris Brown, convicted classified document leaker Chelsea Manning, anti-vaxxer Kent Heckenlively and Gavin
McInnes, founder of the all-male far-right group Proud Boys.[11]

The full-blown media assault on my coauthor had reached its apparent
peak. As if it weren't enough to claim he was a threat to the "good order" of
Australia, he was now lumped together with alleged sexual abuser R. Kelly,
convicted domestic abuser Chris Brown, convicted document thief Chelsea
Manning (although her sentence was commuted by President Obama), and
alleged member of the far-right, Gavin McInnes.

Kent thought it was ridiculously funny, penning a savage satire called
I Fail the Australian Character Test with R. Kelly and Chelsea Manning, in
which he'd join in together with this group in a movie titled *The Dirty Half
Dozen* patterned after any countless number of cheesy World War II films
in which a band of misfits gets together to complete an impossible mission.[12]
In this case, it would be liberating Australia from the grip of the pharmaceutical industry.

But Kent took it all philosophically. "Well, at least I'm still welcome in
six of the world's seven continents," he told me.

Kent always looks on the bright side of life.

CHAPTER ELEVEN

The Way Forward

Yes, I've complained a lot in this book about the sorry state of science these days.

But it's only because we need to have an accurate sense of the problem before we can make significant progress toward solving it.

One of my favorite books and movies is Andy Weir's *The Martian*. In the story, the main character, Mark Watney, an astronaut, is stuck on Mars after his crew believes him to be dead during a sandstorm that threatens their base, and they blast off without him. The actor Matt Damon did a wonderful job of bringing this character to life. I loved how at the end of the movie, Watney is explaining the issue of working intelligently on a challenge to a bunch of astronaut recruits. As he paces around the room, the survivor of an incredible odyssey, he says, "At some point, everything's going to go south on you. Everything's going to go south. And you're going to say, 'This is it. This is how I end.' Now, you can either accept that or you can get to work. That's all it is. You just begin. You do the math. You solve one problem. Then you solve the next one. And then the next. And if you solve enough problems, you get to come home."[1]

I think we can all agree that something has gone south in the health of humanity. And there have been times when I've felt as thoroughly abandoned, as if I were the only astronaut left on an alien world. I could've said, "That's it. This is how I end."

But my faith in God and my faith in science would not allow me to do that.

We simply begin. We get to work. Solve one problem, then the next.

If we solve enough problems, we can enter a golden age of health.

* * *

Let me tell you the reality that exists now for patients.

This is a patient I've been advising for a few years, and his story makes me as angry as anything I've previously shared with you. I will give him the name Sean for purposes of this story, but it is true that he's currently forty-two years old, went to a top ten Ivy League college, and, before the difficulties I'm about to relate, pulled down over three hundred thousand dollars a year as the manager of a group of engineers at a Silicon Valley tech company.

The problem first began in 2004, when he went to the hospital complaining of a fever and chest pains. Tests were unrevealing, and he was given Vicodin and sent home.

In 2015, he had a fever for several days, then began experiencing massive chest pains that he likened to a "massive bowling ball" smashing him in the chest. When he went to the hospital, they immediately found his blood pressure was high, his EKG was abnormal, and they ran a blood test for troponin, which is released into the blood when there's damage to cardiac tissue. Anything over zero means there are problems, while above two means you've suffered a heart attack. Sean's troponin level was over fourteen.

In the intensive care unit, he was diagnosed with acute myocarditis, and in Sean's opinion the team did a heroic job saving his life.

After he had stabilized, a doctor came in and asked to speak to Sean away from his wife.

Sean said anything he had to say his wife could hear, as well.

The doctor then proceeded to ask Sean if he used IV drugs, had recently had a blood transfusion, or engaged in unprotected sex with women other than his wife, such as prostitutes.

Sean laughed at the question. "I'm a boring software engineer. I haven't done any of those things."

The doctor seemed to get a little angry with Sean. "I'm trying to have a real discussion with you, because I think I know what you have. I've seen several cases of acute myocarditis like yours, and all of those patients have AIDS."

Sean's mind swirled at what the doctor was telling him, but none of it made sense. Finally, Sean said, "You did a lot of blood tests. I assume one of them is an HIV-AIDS test. Do you have the results?"

"No, they haven't come back yet, but they should shortly."

"Well, come back and talk to me when you have the results."

A few hours later the doctor returned. Sean did not have HIV-AIDS. The doctor was baffled.

Sean slowly recovered and after a few months tried to return to work but suffered premature ventricular contractions after brief amounts of activity. The doctors ran a series of tests and found he exceeded class-three cardiac stability standards as set by the American Heart Association.

Sean utilized that same relentless drive that had landed him at an Ivy League university and followed him through his previous jobs at five Silicon Valley start-ups. He eventually found an excellent transplant cardiologist who knew a great deal about acute myocarditis, since many patients end up needing a heart transplant.

The transplant cardiologist did a thorough medical review, trying to find something that had been missed, when he came across a troubling finding. "We know the five-year survival rates for somebody who had an attack of acute myocarditis is about fifty percent. But when I looked at your records, it seems like you had something similar, although not as bad, happen in 2004," he told Sean.

"And you don't have any reliable statistics on survival rates for somebody who's had two episodes?"

"Right."

"And if I have a third attack?"

"Yeah, it's a big problem."

The transplant cardiologist was focused on finding answers for Sean, suspecting he had some sort of chronic immune problem that was making him susceptible to these repeated cardiac events. He ran a Lyme test, and it showed five full bands, meeting both New York State and CDC standards for a diagnosis of Lyme disease.

The transplant cardiologist referred Sean to an immunologist with connections to the National Institutes of Health. The cardiologist was interested in getting Sean tested for endogenous viral expression, innate and adaptive immunity, as well as reverse transcriptase testing. Sean knew the immunologist had studied at Columbia University (presumably with Dr. Ian Lipkin), and this was where things started to go off the rails. For the sake of simplicity, let's simply refer to this immunologist as Dr. Bad-Medicine.

Everything seemed to be going fine in the first part of this appointment until Dr. Bad-Medicine said he'd need to punch in some of these tests he was requesting into the National Institutes of Health database.

Dr. Bad-Medicine said he'd be gone for just two minutes but didn't return for about fifteen.

Sean recalls Dr. Bad-Medicine having a little smile on his face when he returned and looking down at the ground, before turning his attention to Sean. "Have you ever heard of CFS, chronic fatigue syndrome?" he asked.

"Yeah."

"I think you have CFS."

Sean replied, "Well, nobody has suggested that. And usually CFS patients have symptoms, but not much in the way of abnormal test readings. I have multiple cardiopulmonary stress tests that show a major biological dysfunction to the point of exceeding long-term disability requirements. I have many blood tests from LabCorp and Quest that show very high immune response to multiple viruses. I have MRIs that show brain-stem inflammation consistent with significant viral activity."

Dr. Bad-Medicine said, "You may have other health problems, but I think CFS is in the mix. Once CFS is in the mix, we don't need to do any more tests because no tests make sense when someone has CFS. Unfortunately, nobody really knows much about CFS."

Sean tried various approaches, such as noting his factor-eight levels for blood clotting being low, or his TGF Beta 1 levels being so high they were consistent with internal hemorrhaging.

"No tests are relevant when somebody has CFS. If you have CFS, that's all you have."

Sean fell back on his math and engineering skills. This was preposterous. "You're telling me that anything plus CFS simply equals CFS? Like an equation?"

"Unfortunately, nobody really knows much about CFS," said Dr. Bad-Medicine.

"So, when did I get CFS? Was it after I was in the ICU for myocarditis? Or I had it before, even though I was working seventy hours a week?"

"It doesn't matter when you got CFS. What matters is you have CFS now."

"That includes blood tests I got when I was eighteen?"

"Once somebody has CFS, none of their past tests really matter."

Sean continued to argue with him. At one point, Dr. Bad-Medicine mentioned there might be clinical trials at the NIH, or he might want to contact Dr. Jose Montoya at Stanford, but he wasn't enthusiastic about any of the prospects.

Sean protested. "I want to be able to enjoy my family, play with my kids, but I can't do that if I keep having these cardiac problems after a few minutes of activity."

"There are no treatments to improve your condition," said Dr. Bad-Medicine. "Since there's no way to get better, there's no reason to do tests of any kind. You will not get the test you want here. You will not get the test you want anywhere. There are no treatments available to help you here. There are no treatments available anywhere. You can see yourself out."

And with that, Dr. Bad-Medicine left the room.

I cannot believe that man is allowed to be a healer.

When I talk so passionately about the attack on me and the science, it's because I know what Sean experienced is a direct result of the plague of corruption that destroyed my reputation. When you murder the truth, you murder the possibility of an answer. We have had the answer for at least a decade! The retroviruses contaminating the vaccines. In cardio-vascular disease, the retrovirus Gary Owens discovered and discussed at the Cleveland Clinic meeting on November 10, 2009, XMRV-2! You see, Gary Owens's expertise is cardiology. What did John Coffin and the NIH do? They made certain Owens's lab did not publish until 2013 and only with Coffin's lab on the paper renaming the Virus BV2. Fraud and cover-up once again. There is nothing resembling truth in medicine in the developed world today!

We must not let this continue.

* * *

The reason Frank Ruscetti went into scientific research rather than becoming a medical doctor was that he believed that discovering basic facts would help promote human health long after this death. This is not as true in the case of physicians. He has always liked the idea of discovering something unknown. I share that passion. His love for scientific research has not changed, but he acknowledges that research has become more difficult because we have more information. Scientists must bring a larger toolbox of techniques to solve the problem.

In order to fix what has gone wrong in science today, Frank has four specific recommendations:

1. Eliminate all financial prizes for scientists, as it hurts cooperation.

2. Eliminate all anonymity for reviewers of grants and publications. If a scientist cannot say something publicly, then why is he saying it at all?

3. Cap the amount of public grant money any individual researcher can receive. The more practicing scientists we have, the better it will be for science in general.

4. Establish a mandatory retirement age to unblock opportunities for the next generation of scientists.[2]

I agree with Frank's proposals to change how we are currently doing science and medicine. I would add repeal of the Bayh-Dole Act, which allows federally funded scientists to patent discoveries. I've always been offended by this act. How can I claim ownership of "intellectual property" when the taxpayers have paid for my education, supplies, equipment, and mentoring? When I entered the field before Bayh-Dole, it seemed as if there were a more collaborative spirit among researchers. Science builds on the work of others. There was no "I" in scientific discovery, only "we." I happily accepted lower pay for intellectual freedom and the opportunity to advance human understanding.

I've always believed scientists to be the best among us, in part because most, like me, didn't go into it for the money. I always believed the most intelligent, creative, and compassionate in research were also the humblest. I do not believe the system is rewarding those qualities in our scientists today. We are turning those good people into outlaws.

* * *

Okay, so that's how Frank thinks we should fix science.

Now, how do we help those who are currently suffering? We have a few ideas.

If we're under constant attack by viruses and retroviruses, not to mention the assault of chemicals that disrupt the proper functioning of our immune system (and don't even get me started about the negative effects of electromagnetic radiation from cell phones and cell towers), what are we to do if we want to regain our health?

We've mentioned many times in this book how many scientific articles we've read in an attempt to understand what's really going on with our health. And it stands to reason that anything we say here is the best plan I have as of 2019. Who knows what will happen in 2020 or the years beyond? However, we believe the principals we're expressing are sound. We're always ready to try something new if it seems promising and is safe. This is a work in progress, but we think at least the outlines of it will hold up five, ten, or fifteen years from now.

* * *

Among the first approaches we might try for any patient we suspect to be suffering from acquired immune dysfunction associated with retroviruses is something called deuterium-depleted water. I am indebted to Dr. Petra Dorfsman, who introduced me to this information, as she has been a pioneer in its use in the United States and supports my health with this amazingly simple solution.

I never thought we'd start with something as simple as water. Everybody knows that the chemical formula for water is two hydrogens and an oxygen (H_2O). The claim is that deuterium, an isotope of hydrogen (D_2O), is clogging up our mitochondria, lowering its output. Our cells then become loaded with deuterium. Petra really got our attention when she told us that deuterium acts as a growth and transforming factor in all microbes, including retroviruses and the cells in which these microbes live. Importantly, deuterium is a cancer-causing oncoisotype in any mammalian cell with mitochondria. Now, remember, we started in science as cancer researchers, so when we investigated deuterium-depleted water, we discovered our interest piqued when we found it was being used in medical settings as an accompaniment to traditional chemotherapy.

The main medical expert putting forth this theory is Dr. Laszlo Boros of Hungary, and this is from the abstract of a 2015 study that Dr. Boros conducted in combination with the University of California, Los Angeles, University of Arizona, and Johns Hopkins University:

> The naturally occurring isotope of hydrogen, deuterium, could have an important biological role. Deuterium depleted water delays tumor progression in mice, dogs, cats, and humans . . . A model is proposed that emphasizes the terminal complex of mitochondrial electron transport chain reducing molecular oxygen to deuterium depleted water (DDW); this affects glucogenesis as well as fatty acid oxidation . . . DDW is proposed her to link cancer prevention and treatment using natural ketogenic diets, low deuterium drinking water, as well as DDW production as the mitochondrial downstream mechanism of targeted anti-cancer drugs such as Avastin and Glivec. The role of deuterium in biology is a potential missing link to the elusive cancer puzzle

seemingly correlated with cancer epidemiology in western populations as a result of excessive deuterium loading from processed carbohydrate intake in place of natural fat consumption.[3]

If there's one thing that's characterized my career, it's being open to new ideas and seeing how they may fit with what we already know, or think we know. I was taught by Frank to integrate new information so we might come to a better understanding of the problem. I'd like to think this type of collaborative thinking in science is widespread, but it's not. I can't tell you how often I've seen a new piece of information presented, real breakthroughs, then watched how a leading scientist who has even a moderately different approach won't even consider it. In my opinion, there's no place for ego in science.

When Petra told me that deuterium acts as a growth factor for microbes including retroviruses, it was like lightning in my brain. If we used that as a working assumption, so many pieces of the puzzle would fall into place. Yes, the use of animal tissue for vaccines and other medical products was certainly a piece of the puzzle, but then when you add this deuterium-rich water that affects the working of the mitochondria, a number of other interesting paths open up. I number myself one of those people who believes that cancer is always in us at some level, but it's the proper functioning of our immune system that keeps it at bay.

What are some other lines of evidence to suggest deuterium-depleted water might be a useful weapon in our arsenal to promote health? We already know that the eating of too much processed food is linked to poor health. Petra and others believe processed food is likely to contain high amounts of deuterium-heavy water. Sources of deuterium-depleted water are likely to be fresh food, especially healthy fats. That's one of the reasons that many people with epilepsy have found success with the so-called ketogenic diet. The ketogenic diet forces the body to take energy from healthy fats. These healthy fats are rich with deuterium-depleted water. That also may be why many find benefits from fasting because it forces our body to burn our fat stores. This causes the release of deuterium-depleted water from within our own fat cells, fueling healthy mitochondria.

This likely explains the health benefits people get from healthy fats. They have the right kind of water in them. And it turns on its head the whole campaign for low-fat foods. We didn't evolve by eating low-fat meals. We are putting unnatural foods into our body in the belief that the type of things we've eaten for millions of years are harming us. It's madness. The other day I saw my husband, David, about to pour some low-fat half and half into his coffee. I grabbed it and tossed it down the drain, saying, "What do you think you're doing?" The poor guy was so surprised. I said, "Look at the first ingredient: high fructose corn syrup. Honey-bear, please eat the full-fat foods. They taste better and they're healthier for you."

I can also add a personal testimony to this story. Prior to my investigation of XMRV, I was tested for exposure to HIV semiannually as I worked with it in the lab. In the XMRV investigation I was a negative control, as were other lab workers. Suddenly in 2011, Frank called me with the news that patient #2623 was testing positive for XMRV with serology and protein assays. Patient #2623 is not a patient; it was me! I had done all my work under Bio-Safety Level 2+ conditions. We wear double gloves, autoclave all trash, and bleach everything. If I had become infected with XMRV, there was only one possible way I had contracted it. The virus was aerosolized, meaning it could float through the air. Now, like Madame Curie, who studied radiation and eventually died from radiation poisoning, I have become a victim of the monster I sought to study.

Those with ME/CFS often complain of foggy thinking, multiple sclerosis-like inability to walk or talk, and I have suffered such bouts as well, especially after long plane flights, which have been shown to create oxidative stress. Yes, I now suffer many of the same problems of the patients I study. The answer to the question of how to treat these conditions is no longer academic to me.

It is personal.

Petra suggested I try the deuterium-depleted water, and I said that sounded great. I'm always willing to try something new if it's safe, and what could be safer than drinking water? For me, the results were remarkable. I was my old self. I literally have the cognitive function I used to have before the last decade of working with the most transmissible of human disease-associated retrovirus families, the XMRVs. If I were to look at my mitochondria, I imagine I'd see them generating the right amount of energy, rather than the danger signals they would when they are under stress.

I think there's also another possible benefit to deuterium-depleted water, and I want to make sure I explain it correctly. If we have the water molecule with the correct isotope of hydrogen, the molecule will actually look different. This correct shape of water means it will be more easily able to penetrate the cells and hydrate them. The mitochondria will then switch from sending out danger signals to providing the energy needed by the cells to do their job of recovery and repair. If the power plant of the body is running efficiently, all the other organs of the body can do their work knowing they have sufficient energy. If there isn't enough energy, the organs of the body, like the foreman of a production facility, must figure out shortcuts to keep things running. Maybe some parts of the plant will need to be shut down in the hope that not too much damage will take place until the power is restored. Deuterium-depleted water can be found online at sites like Amazon.

This may sound heretical if you've only studied the old-time theories of genes and how they work. But we've discovered there's a literal symphony between our environment and our genes.

The official name of this branch of science is "Epigenetics."

Simply put, it means that your environment affects the functioning of your genes. DNA methylation is a primary epigenetic way to turn on and off genes. But it is only one of many. If you're living in a polluted environment, some genes will switch on and others will switch off. The same has also been found happening when people are under high amounts of stress. Certain genes will turn on and others will turn off.

It's for these reasons, among others, that I fear what may come from genetically modified organisms (GMOs) in our food supply. We literally have no idea what these GMOs may do to our interior environment and which genes may inadvertently be turned on or off. I'm a big believer in science studying natural processes, then trying to follow that as much as possible.

* * *

Even though I avoided recreational marijuana my entire life, I am now an enthusiastic backer of medical cannabis for chronic health conditions. My coauthor, Kent, laughs at how the two of us are such squares, as his first legal job was working for the Drug Task Force of the United States Attorney's Office in San Francisco, listening to wiretaps of Oakland drug lords. Now we are both strong backers of medical cannabis.

I like to say we think cannabis has curative properties, meaning formulations of cannabis can provide some assistance to those with chronic health conditions, but for the vast majority it will not be a "cure." I compare what cannabis does in the body to a dimmer switch on neuroinflammation. If your problem is relatively mild, cannabis can certainly seem like a cure, but that's because you didn't really have that far to go to restore immune endocannabinoid balance.

It might be helpful for you to understand cannabis and what we call the endocannabinoid system. Although only discovered in 1987, I believe the endocannabinoid system is one of the most important systems in the body. The endocannabinoid system has been called the "supercomputer that regulates homeostasis in the body." It consists of a group of molecules known as cannabinoids and the receptors to which they bind. We've learned that these receptors regulate a variety of functions, including appetite, pain, inflammation, thermoregulation, muscle control, metabolism, and even controls our response to stress. Professor Raphael Mechoulam, widely acknowledged as the father of cannabis medicine, has gone so far as to state that the "Endocannabinoid system is involved in essentially all disease states."[4] Cannabis has been found to have anticonvulsant, antipsychotic, anti-inflammatory, antioxidant, and antidepressant properties, among others.

When our bodies are under stress, we can't make our natural cannabinoids. This is certainly true of those with ME/CFS and children with autism, as well as those with cancer. It might surprise you to discover that besides hemp and cannabis, the only other rich source of naturally occurring cannabinoids can be found in mother's milk. Yes, that's right, mother's milk has cannabinoids. How many wonderful things can we say about what Mother Nature has provided us in mother's milk?

I believe that with the findings that cannabinoid receptors are present in the brain and stem cells, we have an explanation for why cannabis has therapeutic benefits for many chronic illnesses. Even small doses of plant-derived cannabinoids can signal the body to make more of its own cannabinoids and can also stimulate the building of more cannabinoid receptors. A functional endocannabinoid system is essential for our health.

Because of censorship and corruption in the science of cannabis, many of the studies have been hidden or misrepresented to the public and medical community. Like any natural product made into a therapeutic, it needs the expertise and commitment of scientists in order to translate unbiased science into useful therapeutics.

However, we think formulations of cannabis should be one of the first things that a practitioner uses and can help in a wide variety of situations as a low-cost and remarkably safe intervention. Our job as translational scientists is to quickly pursue these safe, supportive therapies free from political bias. This has not happened to date.

For example, I'm sure you've heard about the water crisis in Flint, Michigan, where high levels of lead were allowed to be in the drinking water. When that story broke, I said, "Just throw the cannabis in the water. It's a

natural heavy metal detoxifying plant. Just don't use it after that. Purify it."
I'm talking about an extract of the whole plant that has been purified. It's
a huge detoxifier, and it would pull out all the heavy metals. There are very
simple, cheap, and effective ways to pull these heavy metals out of people.
Once we know what the enemy is, we can devise strategies to fight it.

The other benefit of cannabis is that the endocannabinoid system is a
lipid-signaling system that is critical to the function of hematopoietic stem
cells and the mesenchymal stem cells, which means we're talking about
the proper functioning of the brain, immune system, digestive system, and
heart. It probably helps to think of your body as one gigantic signaling system. The mitochondria are at the heart of this signaling system because if
you don't have the proper amount of energy, all the organs of the body must
compensate by shutting down or running at reduced power and function.
One of the things we're trying to understand is what the biological markers
of hibernating bears look like. We expect many of them will resemble the
markers of those with ME/CFS and other neuroimmune diseases.

One of the truly horrifying things I've heard over the years related to
energy depletion is how quickly children with autism who die go into rigor
mortis. Think about what that really means. Their bodies are trying so des-
perately to stay alive that when they give up the ghost, rigor mortis sets
in very quickly. I saw this same phenomenon with my aunt who died of
pancreatic cancer in 2011. Chronic disease means your body is trying to
stay alive. You are so close to death that rigor mortis doesn't take long. How
horrifying.

I was shocked at how quickly this happened in my aunt and didn't
really understand it until a researcher, Dr. Robert Naviaux of the University
of California, San Diego, explained it to me. "Chaos is entropy," he said.
"No movement." We think of chaos as erratic movement, so it stunned me
to think of entropy as randomness or stillness. I thought not of the autism
kids, but of the ME/CFS patients who needed to lie in darkened rooms for
most of the day and not move. No light or sound. Any energy they expended
came at the cost of terrible suffering. What a horrible existence to be on the
very edge of death, and you can't move in either direction. I believe there is
a hell after this life, but that certainly sounds like a living hell to me.

There's another way in which recent discoveries suggest we can heal
neurodegenerative diseases and brain tumors. I like to think of it as "brain
washing," but I mean it in a good way.

For decades, we've known of the body's lymphatic system, which is
essentially a fluid superhighway, a protein-rich fluid that washes away the

things your body wants to get rid of, such as bacteria, cancer cells, dead cells, and toxins. A simple way to explain the lymphatic system is to say it's all the fluid in your body that isn't blood. There is a network of connectors and pumps that contain trafficking molecules for immune cells, primarily macrophages (I like to imagine them as little garbage trucks) that gobble up all the bad stuff and safely dispose of it.

There's been a widely held belief that the brain was "immune-privileged," meaning antigens introduced could not elicit an immune response or create inflammation in the brain. Then in 2015, it was discovered that yes, the brain did have its own lymphatic system. Here's how it was reported in the pages of *Neuroscience News*:

> In a stunning discovery that overturns decades of textbook teaching, research-ers at the University of Virginia School of Medicine have determined that the brain is directly connected to the immune system by vessels previously thought not to exist. That such vessels could have escaped detection when the lymphatic system has been so thoroughly mapped throughout the body is surprising on its own, but the true significance of the discovery lies in the effects it could have on the study and treatment of neurological diseases rang-ing from autism to Alzheimer's disease to multiple sclerosis . . .
>
> The unexpected presence of the lymphatic vessels raises a tremendous number of questions that now need answers, both about the workings of the brain and the diseases that plague it.[5]

This is enormously important because while it's been widely acknowledged that vaccines can affect the workings of the immune system in a negative fashion, it was believed they could not affect anything in the brain. The thought was that the brain was off-limits and protected. This discovery says simply that in fact the brain is connected to the immune system, so any-thing that causes a reaction in the immune system will also affect the work-ings of the brain.

This is what you call a big "oops!" and a refutation of the idea that "sci-ence is settled" on anything.

By quieting the inflammation in the nervous system, assisting in cel-lular signaling, and restoring the balance of the endocannabinoid system, cannabis can play a vital role in recovery for many individuals.

* * *

In the previous section on cannabis, I mentioned the observations of Dr. Robert Naviaux, a researcher from the University of California, San Diego, whom I consider to be one of the best thinkers about autism and possible routes to recovery.

Naviaux hypothesizes our modern world is triggering an ancient cellular defense system known as the "cell danger response." He compares it to a castle in the Middle Ages that was designed to be an impregnable fortress, used by the locals in times of emergency. When invaders came and could not be defeated by the King's forces, the villagers would run to the castle, the drawbridges would be raised, and the population could remain there in safety until the threat had passed. Naviaux believes our cells utilize a similar strategy when confronted by pathogens. However, what would happen if the threat remained persistent? The villagers would never come out of the castle. Normal commerce could not resume. While among humans there would be starvation, and likely cannibalism, in such a scenario, for the cell, it would need to go into a state of virtual hibernation.

An article about Naviaux's research from *Science Daily* on March 13, 2013, describes the main points of his theory:

> "When cells are exposed to classical forms of danger, such as a virus, infection, or a toxic environmental substance, a defense mechanism is activated," Naviaux explained. "This results in a change to metabolism and gene expression and reduces the communication between neighboring cells. Simply put, when cells stop talking to each other, children stop talking."[6]

That really covers a lot of ground doesn't it? What we like about the Naviaux theory is it frees us from the narrow question of what substance is causing the problem to the broader question of how damage is being inflicted on the body. We need a clear marker so we can answer the questions of whether the body is in the cell danger response and what we might do in order to send the all-clear signal to the body.

Not to get too geeky on you, but Naviaux identified seventeen inflammatory signaling molecules that identify that one's mitochondria is in the stressed state of the cell danger response. They are called "mitokines," and Naviaux measured them with a specially programmed mass spectrometer in his laboratory. He identified seventeen of these mitokines and had a clear target for his intervention.

Naviaux also identified a promising drug that was likely to quiet the cell danger response. It was called suramin and had been used safely for

more than a hundred years to treat African Sleeping Sickness. I was familiar with suramin because I'd worked with it in the early cancer and AIDS
drug trials, and for some patients it produced fabulous results. At the time
we couldn't discern a pattern, but I've since come to believe that the HIV-
AIDS patients who died the most quickly were infected with both HIV and
XMRV. Naviaux first induced this cell danger in mice (using an injection
that made the immune system of the mice believe they had a viral infection), then tested them with suramin. As reported in *Science Daily*:

> The drug restored 17 types of multi-system abnormalities including brain syn-
> apse structure, cell to cell signaling, social behavior, motor coordination and
> normalizing mitochondrial metabolism. "The striking effectiveness shown in
> this study using APT [anti-purinergic therapy] to reprogram the cell dan-
> ger response and reduce inflammation showcases an opportunity to develop
> an entirely new class of anti-inflammatory drugs to treat autism and several
> other disorders," Naviaux said.[7]

Naviaux would add onto these initial mice studies mice who had a human
age equivalence of thirty years and mice with Fragile X Syndrome, often
referred to as genetic autism. We think of autism as a disease of accelerated
aging, so the genetic diseases like Fragile-X and Rhett's Syndrome molecular pathways may have relevant similarities. (Rhett's Syndrome is caused by
a genetic defect in the DNA methyl-binding protein, MECP2.) The results
were similarly successful in the other two trials, showing that even age or
genetic status might not affect the success of this drug intervention.

Eventually, Naviaux was given permission to try this century-old
African Sleeping Sickness drug on children with autism. In May 2015, the
UCSD Suramin Autism Treatment Trial was started, and it was completed
in March of 2016. The study was a blind study, with five of the children
receiving the drug and five receiving a placebo. In a clinical trial update
from January 18, 2016, Naviaux wrote:

> Suramin produced improvements in all of the core symptoms of autism. All 5
> of the 5 children we think received the treatment showed significant improve-
> ments in language, social interaction, and expression of interest. Suramin
> appears to be working at a fundamental level. By removing the developmental
> barriers caused by the cell danger response, suramin permitted children with
> autism to start moving through developmental stages they had not completed
> before. Physiological abnormalities controlled by the brainstem and related

to low parasympathetic tone are well-known in autism. Many of these were corrected within hours of a single dose of suramin.[8]

This is the parental report from subject number one of the UCSD Suramin trial. I want you to imagine how you would react if this were your child. The parents of this child are both physicians, so one would generally trust their observations:

> Can you imagine being the parent of one of the locked-in patients in the movie "Awakenings"? Our son is 11 years old. At one-and-a-half years old, he knew his colors, shapes and numbers in English, Spanish and Farsi. He could put puzzles together faster than anyone at his age. He was social and engaged others. By three years old all of that language and communication was lost, and he was diagnosed with autism.
>
> The first night, after eating all his dinner, he calmly looked up at me and said, "I finished my dinner." He had never said this before. Usually he would say, "I did it," when he finished his dinner. Two days after the infusion, at his follow-up visit to the infusion center, he said clearly to the nurse, "I want to go to the bathroom again." This was probably the longest sentence of his life. Six days after the infusion, he started asking to try new and different foods. He ate lettuce! This was a big change because he is usually a very picky eater.
>
> We have been working hard, trying every new treatment for autism that seemed reasonable for the past eight years. Nothing has come close to the effects we saw from suramin. Just one dose of suramin gave us a glimpse of that child who was locked in by his disease.
>
> So, can you imagine being the parent of that locked-in patient in the movie "Awakenings"? When the patients finally wake up and smile and dance. But then it's all taken away?[9]

Knowing what we do about the suffering endured by children with autism and their parents, it's difficult for me to read that account and not ask why the federal government isn't moving forward with all possible speed to get this drug approved for use in autism.

How difficult can it be?

Suramin has already been in use for more than a century and is on the World Health Organization's (WHO) list of essential medicines. These are medicines considered to be most effective and safe to meet the important needs in a health system. The only reason there was a problem was that in the 1980s the cancer cowboys gave such high doses to HIV-AIDS and cancer patients that there was kidney damage. The amount we're talking about giving to these children is less than 5 percent of the dose that was given safely for more than seventy-five years.

Perhaps the real problem is suramin is off-patent, which means the company that owns it, Bayer, needs to create a slightly different molecule so that they'll be able to market it and make billions.

Early on in his work, Naviaux was happy to consult with me, but when he found out what had happened to me, he severed all communications. After all, suramin worked best on gamma-retroviruses and the methylation and mitochondrial defects associated with those most susceptible to injury. Still, I'm not sure that when I meet God, I want to tell Him I was safe while people were suffering. If others are suffering, I do not want God to ask me why I valued my own reputation over the needs of others.

The statement below is from an article Naviaux published in 2018. I wish that all scientists could pursue their work without fear of persecution. But I fear that is not the case when one comes up against powerful interests.

> Myalgic encephalomyelitis/chronic fatigue syndrome (ME/CFS) is an energy conservation program—a suite of metabolic and gene expression changes—that permits persistence under harsh environmental conditions at the expense of reduced functional capacity, chronic suffering and disability . . . However, several energy conservation states are known that are activated by harsh environmental conditions. One of these is called *dauer*, the German word for persistence, or to endure. When *dauer* is triggered by harsh conditions, the life expectancy of a classical genetic model system, the 1 mm long worm *Caenorhabditis elegans*, is extended from 2 to 3 weeks to up to 4 months . . . The good news is that the *dauer* state in the worm model is completely reversible. If *dauer* is a good model for ME/CFS, then there is hope that by studying the molecular controls of the *dauer* phenotype, new treatments might be discovered rationally to help stimulate the exit from the *dauer*-like state and begin the process of recovery.[10]

Do you appreciate why we want to study hibernating bears and their physiology? They're certainly a lot closer to humans than a worm. It seems that what we're missing is a key piece of the puzzle, and perhaps suramin, a sponge that soaks up danger signals, is some way to signal to the body that things are fine. The drawbridge can be lowered, and recovery can begin.

If we can get the mitochondria working at its proper level with deuterium-depleted water, tamp down inflammation with cannabis, and then send the all-clear signal with a compound like suramin, we might be setting the stage to heal even our most challenging chronic diseases.

* * *

In 1998, Dr. Andrew Wakefield and twelve colleagues published a case series report in *The Lancet* describing mild to moderate inflammation of the large intestine, as well as swelling of the lymph glands in the intestinal lining of the small intestine in twelve children with autism spectrum disorder. Nine of the parents noted the appearance of BOTH the gastrointestinal problems and autistic behaviors in conjunction with a measles-mumps-rubella (MMR) shot. This is the first known scientific description of what would later come to be called the "gut-brain" connection in autism. Whatever you may think of Andrew Wakefield, it is one of the most important contributions to autism research.

From my background in HIV-AIDS research, it made perfect sense. Retroviruses love to hang out in the digestive tract. A great deal of the immune system is found in the digestive tract. So, if you're a retrovirus and you want to disable the immune system, the gut is where you want to go. Not hard to make that connection, right? That's not rocket science. It's just common sense.

If these retroviruses are affecting the immune system of the gut (and by extension the brain), you'd also expect to find abnormal bacterial populations, as well. This has become a fascinating area of scientific research, and these bacterial populations in the digestive system have become popularly known as the "microbiome."

Probably one of the best thinkers in this area today is Dr. James Adams of Arizona State University. He'd been intrigued by reports that when children with autism and gut problems were treated with an antibiotic, vancomycin, both their autism and gut problems improved, sometimes dramatically. However, when treatment was discontinued, the children relapsed. Adams theorized that the children had an abnormal microbiome, and these bacterial populations were not creating the chemicals necessary for the proper functioning of the body and mind. It might make sense to treat first with vancomycin, then provide material from a person with a healthy microbiome. This material would come from purified human feces. Feces are about 50 percent bacteria. A 2019 article from *Smithsonian* highlighted this area of research, as well as the work of Dr. Adams:

More than a decade ago, a theory began to emerge of a gut-brain connection, where a dysfunction in the gut could also affect brain activity. "Seventy percent of our nerves that go into the central nervous system go into our gut. Why is that?" asks Sarkis Mazmanian, a medical researcher at the California Institute of Technology. Mazmanian noted that in germ-free mice, with no bacteria in the gut, "things like anxiety, locomotion, depression and even brain development seemed to be altered" compared with normal animals.

Mazmanian and a team of researchers demonstrated this gut-brain connection in a mouse model of autism in 2013. Three years later, the team did the same for Parkinson's disease. And recently they showed that transplanting feces from a person with autism into germ-free mice would produce many symptoms of ASD in the animals. [11]

Adams and Mazmanian have Wakefield to thank for that "gut-brain" connection in autism. Mazmanian, though, seems to have done a good job in furthering the research. They and others have found significantly reduced gut varieties of gut bacteria in ASD, ME/CFS among other chronic neuroimmune diseases. Although the germ-free mice are not an exact fit for humans, it's interesting to note that those who don't have any bacteria in their gut have problems with anxiety, locomotion, depression, and altered brain development. The implications of these findings are enormous.

Could anxiety, depression, and other mental health issues be linked to the gut microbiome? Transplanting bacteria from a person with autism into a germ-free mouse makes that mouse display many of the behaviors of autism. This is a remarkable proof of concept. And it hints we're just seeing the tip of the iceberg of a potential scientific revolution. Could we also put Parkinson's disease into the category of affliction, which is influenced, if not caused, by abnormal gut bacteria?

All of this got Dr. Adams thinking about whether first clearing out the gut with an antibiotic, then attempting to recolonize the gut with bacteria from a healthy donor, might change the clinical picture of autism. Adams prepared a small clinical trial. A later passage from the *Smithsonian* article read:

The study enrolled 18 children, ages 7 to 18, with a diagnosis of ASD and significant GI [gastro-intestinal] problems. The regimen was exacting but relatively kid friendly.

They used antibiotics to kill specific classes of bacteria in the gut to mimic the losses in ASD and ME/CFS. The results of the study were exactly as predicted by earlier studies and Dr. Andrew Wakefield.

> 16 kids had at least a 70 percent improvement in their GI symptoms, and importantly, they showed improvements in their behavioral symptoms of autism. That paper was published in January 2017 and so impressed the Pentagon that the Department of Defense agreed to fund a large study of microbial transplants in adults with autism, which began enrolling patients in early 2018.[12]

Further on in the article, they note that two years after treatment, the symptoms had decreased in severity by 47 percent. And while at the beginning 83 percent of the patients were rated as being on the severe end of the autism spectrum, after two years that number dropped to 17 percent with 44 percent having improved so much that they no longer qualified as having autism.

Regardless of your opinion of the cause of autism and many of the other chronic mental and neurological problems in our country, this is hopeful news.

There are caveats, to be certain, such as making sure that no harmful bacteria are cultured in the mixture provided to the patients. I also think it's necessary to make sure that a patient's "leaky gut" is fixed, meaning there's no perforations in the intestinal lining, like what Wakefield first found. "Leaky gut" is now alleged to occur in many different conditions, which is another reason why I think that even if we're talking about autism or ME/CFS, the lessons are probably applicable to many different diseases.

However, there's one thing that concerns me even more. It's the last sentence I quote where it says the paper "so impressed the Pentagon that the Department of Defense agreed to fund a large study of microbial transplants in adults with autism . . ."

The Pentagon?

The Department of Defense?

Why are military organizations taking the lead on this issue? Aren't they involved in building planes, ships, submarines, tanks, bombs, and missiles?

Shouldn't it be the National Institutes of Health? The Centers for Disease Control? Last time I checked, the Pentagon was in charge of killing bad guys, not healing the sick.

Or is there another agenda at work here? Military families must submit to vaccinations, and many have claimed the children of service members have rates of autism and other neurological conditions that are much higher than those of the general public. Perhaps our military is trying to fix the daily battle so many of their families face in their homes, so their members will be able to fight the battles we wage in distant lands.

* * *

Over the weekend of March 22 and 23, 2019, I spoke at a conference in Florida. It was put on by a remarkable young scientist, Dr. Kristin Comella, who reminded me a great deal of myself when I was her age. She has a strong sense of service and has worked herself to the cutting edge of stem cell science. And of course, she'd been attacked by the government with lawsuits and threats to destroy her career but has so far fought them off. What's her crime? According to the FDA, taking your own fats cells out of your body through liposuction and removing the stem cells and reinjecting them into another body part constitutes the administration of a drug.

Dr. Comella is the chief scientific officer of U.S. Stem Cell and the president of the Academy of Regenerative Practices. The conference was the annual meeting of the Academy of Regenerative Practices, and it was a great opportunity to talk with leading figures in the field and compare notes. This is where I always love to be, on the very frontiers of knowledge.

I was joined by other renegades at the conference, including Dr. Andrew Wakefield. He gave a great talk about measles and how the effort to eradicate it was simply driving the creation of new strains and leaving death in its wake. We got the opportunity to sit down and have a long conversation, something we hadn't ever really had, even though we've known each other for a decade. I was able to share with him how Dr. Ian Lipkin of Columbia University had used some of the same scientific sleight of hand to discredit our work with XMRV, as Lipkin had done with his work with the MMR shot and the accompanying gastrointestinal disorders and autism. I told him I couldn't watch the documentary featuring him, *The Pathological Optimist*, because I knew in advance the legal system would be manipulated on the jurisdictional question so he wouldn't be able to have a single day in court to try and clear his name. The same tricks had been played on me in Nevada. I don't think he'd been aware of how similar our two cases were, and when we parted, he told me he wanted to interview me for his next documentary.

I often joke about how nonpolitical I am, so when there was a section of the conference on censorship and they said Roger Stone was speaking, I asked my friend Lori quietly, "Who's Roger Stone?" Lori gave me Roger's backstory. Chairman of Donald Trump's first political campaign, longtime advisor to Trump, political provocateur, cohost of a show called *The War Room* on *Infowars*, and the man most responsible for President George W. Bush winning that fight in Florida against Al Gore. Lori told me how Stone had recently attracted the attention of Special Counsel Robert Mueller's investigation of Russian collusion, when Stone's aggressive reporting made Mueller believe Stone was an actual Russian agent. This sixty-seven-year-old political provocateur who is also well known for his annual Best-Dressed list, found his home raided by the FBI.

When Lori filled me in on the story of Roger Stone, I wasn't sure I wanted to listen to his talk. "Oh, that sounds familiar," I remarked. Yes, I'd also been treated like a fugitive drug lord or potential terrorist. I'd been fifty-three years old when my home was raided, and although I didn't have a CNN camera crew recording my arrest, the powerful forces made sure my mug shot was printed in the journal *Science* as a warning to any researcher who dared associate retroviruses with ME/CFS. "I think this talk might give me PTSD all over again."

I listened to his talk anyway. The stress of what Roger had been going through showed during his presentation and he said he was also fighting a cold. At the end of his presentation, he said he'd take some questions, so I stood and waited to be called on.

"I wanted to ask a question about vaccines and our religious freedoms. When is Trump going to stop them from censoring our science that shows how the health of our country is being destroyed? When is somebody going to start talking about that?" I was unable to control my emotions, and my voice was shaking.

"That's a great question. It's a second-term priority," Stone replied.

After the talk, Lori introduced me to Roger, and she briefly explained what I'd been through. "Hang in there," he said. "We'll get there."

* * *

I am now among the villains according to the media. But like Andy Wakefield said when asked how he has survived to move forward and do great works, it's all about "The simple realization that it's not about me."

The only people who really understand what I've been through are the doctor (Wakefield) who supposedly wants children to die and the political provocateur (Stone) who's allegedly a Russian agent. Though my husband,

David, or Frank Ruscetti might not fully understand the depth of the corruption of our science and medicine, they know me.

I'm blessed to have the opportunity to work alongside Robert Kennedy, Jr. and his Children's Health Defense organization, which is winning remarkable victories against corruption in medicine. Kennedy recently got an admission from the federal government that despite the requirements of the 1986 National Childhood Vaccine Injury Act that a report must be provided to Congress every two years certifying that the vaccine schedule is safe, no such studies have EVER been done.[13]

Think about that.

Your public health agencies have failed to do their job for the past thirty-six years showing the safety of the childhood vaccine schedule. There should be fifteen reports available to every American parent, and Health and Human Services should be almost completed with number sixteen.

There are zero reports testifying as to the safety of the childhood vaccine schedule, or a single vaccine on that schedule. And yet, a WHO essential medication that showed remarkable benefit in a double-blind placebo-controlled study (Suramin) is kept from suffering children!

Who are the real villains?

When I'm in despair, I turn to a section of Robert Kennedy Jr.'s fine book *American Values* in which he recounts his father's love of the Stoic philosophy. Kennedy, Jr. writes:

> That philosophy argued that, in an absurd world, the acceptance of pain, accompanied by the commitment to struggle, transforms the most common men into heroes and provides the most tragic hero peace and contentment. The hero, Sisyphus, condemned by the gods to roll a rock uphill for eternity, only to see it tumble back down, was ultimately a happy man. Even recognizing and accepting the futility of his task, he could find nobility in the struggle. It is neither our position nor our circumstances that define us, according to the Stoics, but our response to those circumstances; when destiny crushes us, small heroic gestures of courage and service can bring us peace and fulfillment. In applying our shoulder to the stone, we give order to a chaotic universe.[14]

I can't tell you I'll submit to this absurd world cheerfully. I will rage against the corruption with all my strength. But I will channel that rage into moving forward and continue to put my shoulder to the stone.

Now that you know my story, perhaps you will be part of the solution. Put your shoulder to the stone with me.

CHAPTER TWELVE

One More Story I Should Probably Tell

Where does all of this leave us?

The plague of corruption is enormous and encompasses many areas of our scientific, medical, and daily political existence.

The pharmaceutical companies have corrupted the laws regarding vaccinations, and a corrupt media has poisoned the mind of the public. The public does not ask the simple question: if vaccines are as safe as sugar water, why do the pharmaceutical companies need to have complete financial immunity and be protected by a battalion of lawyers from the US Department of Justice?

The pharmaceutical companies are the largest contributors to the US Congress and have more lobbyists than all our congressmen and senators combined. That means they control the federal agencies, which is exactly what our constitution with its series of checks and balances was designed to prevent.

The pharmaceutical companies are also bribing the media by their massive advertising buys on the evening news. Don't believe me? Watch an hour of your favorite news station and count the number of pharmaceutical ads. Chances are that 50–60 percent of the ads will be for pharmaceutical companies. Our colleagues are particularly outraged by the commercial of the grandmother turning into a big, bad wolf if she does not get a whooping cough vaccine.

We understand how people can get discouraged when they see such a gauntlet of legal, financial, and political firepower arrayed against us.

But they underestimate those of us who do the hard work of understanding biological processes. We plan for many different contingencies. You see, we're used to dealing with the unknown. We simply hadn't planned on the dark side of human nature that they have shown us as we have tried to answer the most pressing questions of human health.

* * *

Since June 6, 1983, when I started working for Frank, I realized he embodied the very best of science and character. He always wanted to see all the data all the time. Massaging the data with statistics was absolutely unacceptable at weekly lab meetings where everyone was free to offer opinions. That was something he drilled into my head, and I've never forgotten it. It's why, even after I got married to David in 2000 and relocated to the West Coast, we continued to discuss our data or recent publications, usually having a long phone call at 5 a.m. Pacific time and 8 a.m. Eastern time.

Frank became an inside player in science in the 1970s, at a time when the corruption was less blatant and it was possible to go through a career without seeing the darkness. And if you don't see the darkness, it's difficult to believe it exists.

Bernie Poiesz, the MD who isolated the first human retrovirus, HTLV-1, while working as a clinical fellow for Frank, has described Frank as one of the last true Renaissance men, a cross between Leonardo Da Vinci and Rocky Marciano, because he is so widely read and never pulls punches.[1] How were Bernie and Frank able to isolate that first human retrovirus, when the Japanese had been working on the problem for years? Because Frank taught Bernie what he knew about growing viruses in cell cultures and following reverse transcriptase activity. Yes, Robert Gallo ended up getting the Lasker Award (often referred to as the American Nobel Prize for scientific research) for the discovery of HTLV-1. But it was Bernie and Frank who did the work. Perhaps Gallo would have won the Nobel Prize for HIV if he had done the same a few years later?

The credit has gone rightfully to Luc Montagnier, the French researcher who isolated the HIV virus. But Montagnier might never have won his Nobel Prize in 2007 if he had not been wise and learned the lessons of HTLV-1 isolation. Unlike Gallo, who likes to hog all the credit, Montagnier has generously given credit to Frank on multiple occasions for his contributions in HIV-AIDS research.

We would never have been able to isolate XMRVs if it hadn't been for what Frank had taught me. In retrospect, I realize that Frank and I existed in our own little bubble in that Bio-Safety Level 3 lab at the National Cancer Institute, thinking about and publishing all the data, even the data we did not understand.

Frank has been at the forefront of discovery of human disease-associated retroviruses and the function of the immune system because of his vision and his work ethic. One would have thought this would have given him some protection.

But it didn't.

* * *

I believe the campaign against Frank Ruscetti kicked into high gear after the First International Symposium on XMRV, when the response to the critical question of Francis Collins of where we got the control group results of 4 percent from the United Kingdom was from blood bank samples in that country.

It was after that meeting that Collins tasked Fauci to fund and conduct a "confirmation study" of our work. Apparently, the study by Shyh-Ching Lo and Harvey Alter, a Lasker Award-winning scientist, wasn't enough of a confirmation to satisfy them!

Maybe it was a coincidence, but prior to 2010, the National Cancer Institute was run by a fine doctor named John Niederhuber, who was supportive of the work with XMRV and prostate cancer. But what happened in 2010, just as the XMRV freight train was gathering momentum?

They brought in Harold Varmus to take over from John Niederhuber.

Just so you know how absurd this was, you should know that Harold Varmus was director of the National Institutes of Health from 1993 to 1999 in the Bill Clinton administration. The National Cancer Institute is a division of the National Institutes of Health. After leaving as director of the National Institutes of Health, Varmus went to work as the president of the Memorial Sloan Kettering Cancer Center in New York City, where he served for ten years.

Let's talk about downward mobility and see if this makes any sense.

Depending on experience (and probably your political backing), the director of the National Institutes of Health will make somewhere up to $230,000 a year.

In 2016, it was reported that the president and CEO of the Memorial Sloan Kettering Cancer Center, Dr. Craig Thompson, made $2,944,000, or about ten times what one would make as director of the National Institutes of Health.

Now, the latest information on a salary for the director of the National Cancer Institute (a division of the National Institutes of Health) is up to a little more than a $151,000.

Okay, so let's get this chain of events straight.

You're Harold Varmus and you've got both a Nobel Prize and a Lasker Award.

You serve as the director of the National Institutes of Health for eight years.

After serving as director of the National Institutes of Health, you pick up a sweet gig at a New York cancer clinic where you make a couple million dollars a year.

Then suddenly you decide you want to head up a division of the National Institutes of Health (National Cancer Institute), where at most you'll make a little over a hundred and fifty thousand a year, about five percent of what you'd made the previous year.

I just don't buy it.

Harold Varmus was brought on to get Frank Ruscetti under control.

* * *

When Harold Varmus took over at the National Cancer Institute, he directed a team led by John Coffin to discredit Frank Ruscetti.

This was May of 2010, and there was a great deal happening with XMRV. We knew that the Shyh-Ching Lo and Harvey Alter confirmation study was in press in the journal of the *Proceedings of the National Academy of Sciences* (PNAS). That second confirmatory study should have made us bulletproof. The guardians of public health must surely have been on high alert, wondering how they could derail XMRV.

Coffin would go on to have meetings with Frank, usually attended by Steve Hughes and Doug Lowie, a Lasker Award winner for developing the Gardasil vaccine, which is now destroying the lives of so many young girls and boys, often making them sterile. It paralyzes some of them and kills others.

At one particularly contentious meeting, Hughes said to Frank, "This isn't an inquisition," when Frank grumbled why they were asking so many questions about how he was getting all the positives for XMRV and other researchers, as well.

"Sure looks like one," Frank said back to them.

Frank didn't say much to me about those monthly meetings, although he often compared them to being waterboarded in order to get Frank to renounce the raw data. In October of 2010, as the evidence of a heavily contaminated blood supply began to mount, Frank recounted Coffin angrily stating, "*Science* started this, and *Science* is going to end this!"

* * *

After I gave that talk in Ottawa on September 22, 2011, I was burned at the stake as predicted by Coffin. We had integrated our seminal discoveries from HIV-AIDS as well as DNA methylation to show how this new family of retroviruses was devastating humanity. They said it was fraud. Really? Science reviewers and editors reviewed all the data and had them peer-reviewed prior to publication. The journal cut two-thirds of the paper prior to publication, where we detailed the culturing of cells with the DNA methylation inhibitor. Because we refused to be intimidated, they planned to steal all the raw data when I was fired on September 29, 2011, then force the retraction of our paper. The journal *Science* knowingly published fraud September 22, that Michael Busch and Simone Glynn-led "Failure to Confirm" study instead of the true study that the blood supply was contaminated. After I was released from jail in 2011, Frank's punishment was to allow a full investigation of all the data of the XMRV studies, including that figure produced in Frank's lab at the National Cancer Institute.

Of course, Frank's lab passed with flying colors and yet the retraction was forced anyway, and the wording of that retraction essentially states that the figure was fraud. It was not!

Frank and his wife, Sandy, were both forced to retire in 2013, regardless of the facts, just as they had put the XMRV genie back in its bottle and claimed it was all contamination. For good measure, they forced Frank to coauthor another purely fraudulent paper, which blamed all the contamination on my lab in the WPI instead of the NCI lab's in building 535. John Coffin kept his promise. Science ended our careers and the hopes of millions who are suffering from a range of terrible chronic diseases. But don't worry about John Coffin. Coffin was awarded a patent in the summer of 2011 for detection of XMRV contaminants in the blood supply and biological materials.

They'd just about wrapped up everything, and now all that was left to do was get rid of the evidence. My lab in Nevada had been locked down, my notebooks taken, and all the raw data taken. Lipkin had performed the requested "debunking" study, and Silverman at the Cleveland Clinic and Whittemore and the Nevada institutions continue to be awarded millions of dollars of funding. It's no different from the medical staff who got paid off from that first outbreak at the Los Angeles County General Hospital in 1934–1935.

Just continue the cover-up and carry on.

* * *

But what no one at the National Cancer Institute ever realized was that Frank spent his last two years in Frederick, MD, doing critical work. Frank went back to our original samples and reisolated XMRV, obtaining a full sequence. He saved the liquid nitrogen samples of our original stocks of B cell lines and patient samples, which had never touched Building 535, where they'd contaminated a fermenter of our natural isolate of XMRV by placing it next to a fermenter with VP62.

There could be no question of contamination with the original samples saved by Frank.

And when the order came for Frank to clean out his lab, he was told he could take all of his work and materials, except for any records and biological samples regarding XMRV. All samples he was ordered to autoclave and the data to be destroyed Thompson/CDC style. They even sent a security guard to make sure that Frank complied.

Frank threw all the data into a blue recycling bin and packaged up the biological samples and sent them to a biostorage facility in the Midwest, where we've been paying five hundred dollars a month to store them since. To have destroyed the samples or data would have been in violation of federal law.

Frank and Sandy had long planned to retire in Carlsbad, California, and purchased a home there around fifteen years ago, which they often generously offered as a vacation getaway to cancer patients in my support groups. One day, a few weeks after they moved in, Frank showed up at my door carrying the large blue recycling bin.

"I don't know how this shit got in my moving van," he said, pointing to it.

"Thank you," I whispered, tears welling in my eyes.

I looked through the data. They were all there. They may have taken my notebooks and those of my research team, but collaborators are bound to maintain records of ALL of their data. The inquisition had confirmed Frank's lab had maintained accurate records of all data.

I scanned every piece of those data into hard drives, and for good measure we even met with the FBI in the office of my faithful attorney, David Follin, and gave them a copy of the hard drive. If we ever get some honest people at the top of the FBI, maybe we can finally prove the corruption and end this reign of darkness. I know it doesn't bring back those who have already lost their lives to this plague or make up for the suffering of millions today. We have done all we could.

Now, as always, it is up to God. I may be a scientist, but I've always believed in something beyond this world. At some point, we must lay our troubles down and let God be God, working things out in His own mysterious ways. He will put the right people on these challenges, perhaps those with more wisdom and discernment than we who have gone before, and they will lead us to a better world.

Blessings and good health to you all.

Notes

Introduction

1 Byron Marshall Hyde, MD, et al., *The Clinical and Scientific Basis of Myalgic Encephalomyelitis/Chronic Fatigue Syndrome* (Ottawa, Canada: The Nightingale Research Foundation, 1992), 125.
2 John R. Paul, MD, *A History of Poliomyelitis* (New Haven, London: Yale University Press, 1971), 224.
3 Maurice Brodie, "Attempts to Produce Poliomyelitis in Refractory Lab Animals," *Experimental Biology and Medicine* (March 1, 1935), 832–836, doi: 10.3181/00379727-32-7876.
4 W.A. Sawyer et al., "Vaccination Against Yellow Fever with Immune Serum and Virus Fixed for Mice," *Journal of Experimental Medicine* (May 31, 1932), 945–969.
5 John F. Kessel et al., "Use of Serum and the Routine and Experimental Laboratory Findings in the 1934 Poliomyelitis Epidemic," *American Journal of Public Health and the Nation's Health*, Vol. 24, No. 12, (December 1934), 1215–1223. Doi: 10.2105/AJPH.24.12.1215.
6 G. Stuart, "The Problem of Mass Vaccination Against Yellow Fever," *World Health Organization—Expert Committee on Yellow Fever*, September 14–19, 1953, Kampala, Uganda, Presentation.
7 Hillary Johnson, *Osler's Web: Inside the Labyrinth of the Chronic Fatigue Syndrome Epidemic* (New York: Penguin Books, 1996), 200.
8 The Rockefeller Foundation Annual Report (1936) (1937) (1938) (1939).
9 Vincent Lombardi, Francis Ruscetti, Judy Mikovits, et al., "XMRV in Peripheral Blood Cells of Patients with Chronic Fatigue Syndrome," *Science*, Vol. 326, (October 23, 2009), 585–588.
10 Ben Berkout et al., "Of Mice and Men: On the Origin of XMRV," *Frontiers in Microbiology*, Vol. 1, Article 147, (January 17, 2011), 4–5.
11 Ian Lipkin, Public Conference Call with the Centers for Disease Control, September 10, 2013. Transcript by ME/CFS Forums.com/wiki/Lipkin.

CHAPTER ONE

1 Jon Cohen, "The Waning Conflict Over XMRV and Chronic Fatigue Syndrome," *Science*, Vol. 333, September 30, 2011, www.science.sciencemag.org/content/333/6051/1810.summary.

2   Judy Anne Mikovits vs. Adam Garcia et al., Plaintiff's First Amended Complaint and Jury Trial Demand, Case no. 2:14-cv-08909-SWV-PLA, filed July 27, 2015.

CHAPTER TWO

1   Sarah Yang, "Virus in Cattle Linked to Human Breast Cancer," *Berkeley News*, September 15, 2015, www.news.berkeley.edu/2015/09/15/bovine-leukemia-virus-breast-cancer/, accessed March 16, 2019.

CHAPTER THREE

1   Benjamin Wesiser, *A Secret Life: The Polish Officer, His Covert Mission, and the Price He Paid to Save His Country Public Affairs* (New York; Public Affairs, a member of the Perseus Group, 2004).
2   Michael Kellet, *The Murder of Vince Foster* (Columbia, Maryland; CLS Publishers, 1995).
3   Michael Isikoff and Dan Balz, "Foster Note Reveals an Anguished Aide," *Washington Post*, August 11, 1993.
4   Telephone Interview with Thomas and Candace Bradstreet by Kent Heckenlively, May 25, 2016.
5   Kuan Teh-Jeang, "Moving to the Fore," *Retrovirology*, August 13, 2012, doi:10.1186/1742 -4690-9-66.
6   Matthew Cockerill, "Obituary: Kuan-Teh Jeang," *Retrovirology*, March 21, 2013.
7   Krista Delviks-Frankenberry et al., "Generation of Multiple Replication Competent Retroviruses through Recombination between PreXMRV-1 and PreXMRV-2," *Journal of Virology*, (November 2013): doi:10.1128/JVI.01787–13.
8   Baxter Dmitry, "CDC Doctor, Who Claimed Flu Shot Caused Outbreak, Missing Feared Dead," *NewsPunch*, February 22, 2018.
9   Phil McCausland, "CDC Employee Timothy Cunningham Went Missing More Than a Week Ago," *NBC News,* February 23, 2018.
10   Tonya Layman, "Forty Under Forty," *Atlanta Business Chronicle*, October 30, 2017.
11   Tanasia Kenney, "New Report Hints at 'Personal Struggles' Dr. Timothy Cunningham Faced Weeks Before his Suicide," *Atlanta Black Star*, June 6, 2018, www.atlantablackstar .com/2018/06/06/new-report-hints-at-personal-struggles-dr-timothy-cunningham -faced-weeks-before-his-suicide/.
12   Alexis Stevens, "EXCLUSIVE: CDC Researcher Had Personal Struggles Before Suicide," *Atlanta Journal-Constitution,* June 4, 2018.
13   Alex Horton, "The CDC Researcher Who Mysteriously Vanished Had Recently Been Passed Over for a Promotion, Police Say," *Washington Post*, March 1, 2018.
14   Beverly Gage, "What an Uncensored Letter to M.L.K. Reveals," *New York Times*, November 11, 2014.

CHAPTER FOUR

1   Paul Kix, "In the Shadow of Woburn," *Boston* magazine, September 22, 2009.
2   Ibid.
3   Dan Olmsted, "Olmsted on Autism: 1979 Wyeth Memo on DPT," *Age of Autism*, August 12, 2008.

4   Richard Nixon, "Statement on Chemical and Biological Defense Policies and Programs," November 25, 1969, www.2001–2009.state.gov/documents/organization/90920.pdf.
5   Ibid.
6   Ibid.
7   Ibid.
8   Bishop Randy White Biography, www.withoutwalls.org/bishop-randy-white/Accessed December 31, 2018.
9   Cameron Dodd, "Supreme Court Won't Hear $750 Million Fort Detrick Contamination Death Lawsuit," *Frederick News Post*, May 25, 2018.

CHAPTER FIVE

1   Walt Bogdanovich and Eric Koli, "2 Paths of Bayer Drug in 80's: Riskier One Steered Overseas," *New York Times*, May 22, 2003.
2   "Contaminated Blood Scandal: We Are Sorry, Says Government," *BBC News*, September 26, 2018.
3   Andrew Pollack, "Japanese Suits on H.I.V.-Tainted Blood Settled," *New York Times*, March 15, 1996.
4   Phillip Boffey, "Long-Running Debate on AIDS: How Well Did Americans Respond?" *New York Times*, October 13, 1987.
5   Abstract on July 22, 2009 Meeting—"Public Health Implications of XMRV Infection, Center for Cancer Research, Center of Excellence in HIV/AIDS and Cancer Virology (CEHCV)," National Cancer Institute.
6   U.S. Senator Harry Reid, Letter to Dr. Frank Ruscetti, November 17, 2009.
7   Ana Sandoiu, "Alzheimer's May Soon Be Treated with HIV Drugs," *Medical News Today*, November 26, 2018, www.medicalnewstoday.com/articles/323797.php.

CHAPTER SIX

1   "Strains of Xenotropic Murine Leukemia-Related Virus and Methods for Detection Thereof," US Patent and Trademark Application, submitted April 6, 2011, #20110311484.
2   CBS, Suzanne Vernon: 'Agency Heads are Scared to Death . . . if XMRV Works Out,'" *Phoenix Rising Forum*, Discussion in "Action Alerts and Advocacy," February 23, 2011.
3   Shyh Ching Lo, Natalia Pripuzova, et al., "Detection of MLV-Related Virus Gene Sequences in Blood of Patients with Chronic Fatigue Syndrome and Healthy Blood Donors," *Proceedings of the National Academy of Sciences*, August 23, 2010, doi:10.1073/pnas. 1006901107.
4   Judy Mikovits, Frank Ruscetti, et al., "Detection of Infectious XMRV in the Peripheral Blood of Children," First International Conference on XMRV, Abstract, September 7, 2010.
5   Press Release, ORTHO, June 22, 2010.
6   Cerus, 2017 Annual Report, p. 5.
7   Cerus, "Whittemore Peterson Institute and Cerus Announce Inactivation of XMRV in Platelets and Red Blood Cells by the INTERCEPT Blood System," Press Release, September 7, 2010, www.businesswire.com/news/home/20100907005754/en/Whittemore-Peterson-Institute-Cerus-Announce-Inactivation-XMRV.
8   Ibid.

⁹ Cerus, "Cerus Announces Preliminary Fourth Quarter and Full Year 2018 Product Revenue," Press Release, January 7, 2019.

CHAPTER SEVEN

¹ Joseph DeRisi, "Hunting the Next Killer Virus," *TED Talks*, Monterey, California, January 29, 2006.
² Ibid.
³ Ibid.
⁴ Yu-An Zhang, Adi Gazdar, et al., "Frequent Detection of Infectious Xenotropic Murine Leukemia Virus (XMLV) in Human Cultures Established from Mouse Xenografts," *Cancer Biology and Therapuetics*, October 1, 2011, 617–628; doi: 10.4161/cbt.12.7.15955.
⁵ Ibid.
⁶ Jay A. Fishman, Linda Scobie, Yashuiro Takeuchi, "Xenotransplantation-Associated Infectious Risk: A WHO Consultation," *Transplantation*, Vol. 19; 72–81 (2012); doi: 10.1111/j.1399–3089.2012.00693.x.
⁷ Ibid.
⁸ Francis Ruscetti, Judy A. Mikovits, et al., "Development of XMRV Producing B Cell Lines from Lymphomas from Patients with Chronic Fatigue Syndrome," *Retrovirology*, June 6, 2011, 8 (Suppl.1): A230 doi:10.1186/1742-4690-8-S1-A230.
⁹ Frank Ruscetti, email to Robert Silverman, July 7, 2011.

CHAPTER EIGHT

¹ Kent Heckenlively, "Inoculated: How Science Lost its Soul in Autism," *Waterfront Press*, 181, (2016).
² Nora Freeman Engstrom, "A Dose of Reality for Specialized Courts: Lessons From the VICP," *University of Pennsylvania Law Review*, Vol. 163, 1659, (2015).
³ Nora Freeman Engstrom, "Heeding Vaccine Court's Failures," *National Law Journal*, June 29, 2015.
⁴ Telephone Interview with Special Master Gary Golkewicz by Kent Heckenlively, January 29, 2016.
⁵ Ibid.
⁶ "Scientific Review of Vaccine Safety Datalink Information, June 7–8, 2000, Simpsonwood Retreat Center, Norcross, Georgia," Centers for Disease Control and Prevention, Accessed February 7, 2016, National Immunization Program, www.thinktwice.com /simpsonwood.pdf.
⁷ Ibid.
⁸ MARC Inc. Response to Questions to Andrea Heckman at Law Offices of Jeffrey M. Leving, Ltd.
⁹ Ibid.
¹⁰ Ibid.
¹¹ Written Testimony at Georgetown University by Dr. Theresa Deisher, September 2008, www.bioethicsarchive.georgetown.edu/pcbe/transcripts/sept08/deisher_statement.pdf.
¹² Letter to Clifford Shoemaker from Frank Ruscetti, Bhattacharyya v. HHS, No. 16-195V (ECF).
¹³ Expert Opinion Testimony of Judy Mikovits, PhD, and Francis Ruscetti, PhD (McKown, CM).

14 Letter to Clifford Shoemaker by Frank Ruscetti and Judy Mikovits, Re: George Dominguez vs. HHS, Case 12-378V.

15 George Dominguez v. Secretary of Health and Human Services, Published Decision on Remand Awarding Attorneys' Fees and Costs on an Interim Basis; Attorneys' Fees and Costs: Hourly Rate for a PhD Immunologist, May 25, 2018, Special Master Christian J. Moran, No. 12-378V, Case 1:12-vv-00378-EDK.

16 Ibid.

17 Ibid.

18 Richard Smith, "Doctors are Not Scientists," *British Medical Journal*, June 17, 2004, doi .org/10.1136/bmj.328.7454.0-h.

19 Email from Frank Ruscetti to Kent Heckenlively, April 4, 2019.

20 Sharyl Attkisson, "How a Pro-Vaccine Doctor Reopened Debate About Link to Autism," *The Hill*, January 13, 2019.

21 Ibid.

22 Sharyl Attkisson, "Dr. Andrew Zimmerman's Full Affidavit on Alleged Link Between Autism that U.S. Government Covered Up," *Sharyl Attkisson.com*, January 6, 2019, accessed February 24, 2019, www.sharylattkisson.com/2019/01/06/dr-andrew-zimmermans -full-affidavit-on-alleged-link-between-vaccines-and-autism-that-u-s-govt-covered-up/.

23 Ibid.

24 Ibid.

25 Ibid.

26 Letter from Dr. Andrew Zimmerman to attorney Clifford Shoemaker, November 30, 2007.

27 Dr. Richard Kelley, written affidavit, January 24, 2016.

28 J.B. Handley, *How to End the Autism Epidemic* (White River Junction, VT: Chelsea Green Press, 2018), 194.

29 Sharyl Attkisson, "Dr. Andrew Zimmerman's Full Affidavit on Alleged Link Between Autism that U.S. Government Covered Up," *Sharyl Attkisson.com*, January 6, 2019, accessed February 24, 2019, www.sharylattkisson.com/2019/01/06/dr-andrew -zimmermans-full-affidavit-on-alleged-link-between-vaccines-and-autism-that-u-s-govt -covered-up/.

CHAPTER NINE

1 "Zoonotic Diseases," Centers for Disease Control and Prevention Website, www.cdc .gov/onehealth/basics/zoonotic-diseases.html, accessed March 14, 2019.

2 Diana K. Wells, "Zoonosis," Healthline, www.healthline.com/health/zoonosis, accessed March 14, 2019.

3 Edward Hooper, "The Origins of the AIDS Pandemic," *AIDS Origins*, May 25, 2012, www.aidsorigins.com/print/origin-aids-pandemic.com, accessed March 14, 2019.

4 Ibid.

5 David Quammen, *The Chimp and the River: How AIDS Emerged from an African Forest*, (New York: W.W. Norton, 2015), 84–85.

6 Ibid, 130.

7 Edward Hooper, "Edward Hooper Biography," AIDS Origin, www.aidsorigins.com /print/biography, May 26, 2004, accessed March 14, 2019.

8 Annie Jacobsen, *The Pentagon's Brain—An Uncensored History of DARPA, America's Top Secret Military Research Agency*, (New York, Back Bay Books, 2015), 298.

9   Ibid, 298–299.

10 Andrea Lisco, Cristope Vanpouille, and Leonid Margolis, "War and Peace between Microbes: HIV-1 Interactions with Co-infecting Viruses," *Cell Host and Microbe*, November 19, 2009, doi:10.1016/j.chom.2009.10.010.

11 "History of Ebola Virus Disease," Centers for Disease Control and Prevention, November 9, 2018, www.cdc.gov/vhf/ebola/history/chronology.html, accessed March 19, 2019.

12 Ibid.

13 Ibid.

14 Yoichi Shimatsu, "The Ebola Breakout Coincided with UN Vaccine Capaigns," *Rense News*, August 12, 2014, www.rense.com/general96/ebobreakout.html.

15 M. Fernandez-Garcia, M. Majumdar, et al., "Emergence of Vaccine-Derived Polioviruses during Ebola Virus Disease Outbreak, Guinea, 2014–2015," *Emerging Infectious Diseases*, *24*(1), (2018): 65–74. www.dx.doi.org/10.3201/eid2401.171174.

16 Kent and Amber Brantley with Dave Thomas, *Called to Life: How Loving Our Neighbor Led Us into the Heart of the Ebola Epidemic* (Colorado Springs, CO: WaterBrook, 2015), 1.

17 Ibid., 12.

18 Ibid., 80.

19 Ibid, 128.

20 Ethan Siegel and Alex Berezow, "Opting Out of Vaccines Should Opt You Out of American Society," *Scientific American,* March 21, 2019, www.blogs.scientificamerican.com/observations/opting-out-of-vaccines-should-opt-you-out-of-american-society/.

21 Ibid.

22 General Maddow, "BOMBSHELL, Corvela Releases Next Vaccine Analysis Results," *Real News Australia*, December 23, 2018, link to Italian study—www.drive.google.com/file/d/1mufQ9Ueoph4T4BufJ71M6v95KvOfhXAw/view.

CHAPTER TEN

1   Telephone Interview with Dr. Andrew Wakefield by Kent Heckenlively, February 25, 2016.

2   Sophie Borland, "Dishonest and Irresponsible Doctor who Triggered MMR Vaccine Scare is Struck Off," *Daily Mail*, May 24, 2010.

3   Ibid.

4   "Events Surrounding the De Stefano et al (2004) MMR-Autism Study," Prepared by Dr. William Thompson for Congressman William Posey, 2–5, September 9, 2014.

5   Comments of Congressman William Posey, Congressional Record, Vol. 161, No. 121, July 29, 2015.

6   Kent Heckenlively, "Australia Bans Autism Mom as an ENEMY OF THE STATE," *Bolen Report*, August 12, 2015.

7   Kent Heckenlively, "Kent Heckenlively's Dangerous Science Tour Heads to Australia," *Bolen Report*, August 15, 2017. www.bolenreport.com/kent-heckenlivelys-dangerous-science-tour-heads-australia/.

8   Stephanie Pending, "World's Number 1 Anti-Vaxxer Kent Heckenlively Denied Entry to Australia," *Sydney Morning Herald*, August 31, 2017.

9   Notification of Cancellation Under Section 128 of the Migration Act 1958, sent to Kent Heckenlively by Australian Government: Department of Immigration and Border Protection, August 31, 2017.

10   Decision Record of Visa Cancellation Under Section 128 of the Migration Act 1958, sent to Kent Heckenlively by Australian Government, Department of Immigration and Border Protection, August 31, 2017.

11   "R. Kelly Deletes Post About International Tour Amid Backlash," *The Hollywood Reporter*, February 7, 2019.

12   Kent Heckenlively, "I Fail the Australian Character Test," *Bolen Report*, February 10, 2017, www.bolenreport.com/i-fail-the-australian-character-test-with-r-kelly-and-chelsea -manning/.

CHAPTER ELEVEN

1   *The Martian*, 2015, Twentieth Century Fox.

2   Email from Frank Ruscetti to Kent Heckenlively, April 3, 2019.

3   Laslo Boros et al., "Submolecular Regulation of Cell Transformation by Deuterium Depleting Water Exchange Reactions in the Tricarbolic Acid Substrate Cycle," *Medical Hypotheses* (2015) vol. 87, 69–74, www.dx.doi.org/10.1016/j.mehy.2015.11.016.

4   Miles O'Brien Interview of Dr. Raphael Mechoulam, "Medical Marijuana Research Comes Out of the Shadows," *PBS Newshour*, July 13, 2016, www.pbs.org/newshour /show/medical-marijuana-research-comes-shadows.

5   "Researchers Find Missing Link Between the Brain and Immune System," *Neuroscience News*, June 1, 2015, www.neurosciencenews.com/lymphatic-system-brain -neurobiology-2080/.

6   "Drug Treatment Corrects Autism Symptoms in Mouse Model," *Science Daily*, March 13, 2013, www.sciencedaily.com/releases/2013/03/130313182019.htm.

7   Ibid.

8   Robert Naviaux, "Suramin Treatment of Autism-Clinical Trial Update," University of California, San Diego, School of Medicine, January 18, 2016.

9   UC San Diego Health, "Parent Personal Statements of Their Observations from Phase I/II Randomized Clinical Trial of Low Dose Suramin in Autism Spectrum Disorder," accessed April 1, 2019, www.health.ucsd.edu/news/topics/Suramin-Autism/Pages/Parent -Statements.aspx.

10   Robert Naviaux, "Metabolic Features and Regulation of the Healing Cycle—A New Model for Chronic Disease Pathogenesis and Treatment," *Mitochondrion*, August 2, 2018, www.doi.org/10.1016/j.mito.2018.08.001.

11   Bob Roehr, "How the Gut Microbiome Could Provide a New Tool to Treat Autism," *Smithsonian*, June 14, 2019.

12   Ibid.

13   "Stipulated Order Confirming Non-Compliance with 42 USC 300 AA-27C—The 1986 National Childhood vaccine Injury Act," Health and Human Services, July 9, 2018, accessed April 2, 2019, www.icandecide.org/wp-content/uploads/whitepapers /Stipulated%20Order%20copy.pdf.

14   Robert F. Kennedy, Jr., *American Values: Lessons I Learned From My Family* (New York: HarperCollins, 2018), 286–287.

CHAPTER TWELVE

[1] Kendall Smith, "The Discovery of the Interleukin 2 Molecule" (from Dr. Kendall Smith's Immunology Research Site. Dr. Smith is the Rochelle Butler Professor of Medicine and Immunology at Cornell's Weil Medical College and Graduate School of Biomedical Sciences, Accessed August 21, 2019, www.kendallsmith.com/molecule.html).

Acknowledgments

From Judy:
I would like to thank my family for their unwavering support throughout this very difficult decade. My parents Gloria Furr-Fornshill and John L
Mikovits Jr. and my siblings John, Julie, and Karen and their families fought
for me with every resource they had and more importantly never blamed me
for the humiliation and dishonor this plague of corruption brought to our
family name and our proud Cherokee and Austrian/Hungarian heritage.
My mom raised us alone from the age of ten and taught us above all else
integrity and honesty. To mom the most egregious offense was to be silent
about any wrong. To witness a wrong and do nothing generated explosive
anger and guaranteed fierce punishment. As she lay dying earlier this year,
I could get a laugh by reminding her the worst of punishments were usually
precipitated by the statement "God gave you a mouth, USE IT" and in later
years her beloved husband Ken would say to any who would listen "Judy's
mouth gets Judy's body in trouble"; to which mom would beam with pride.
I likely would not have survived without my dear husband David. There
simply is no kinder human being. Few men would willingly lose everything
to keep their wife out of jail. Whenever I get discouraged he texts me two
songs: Marilyn McBride's "Anyway" and Train's "Calling all Angels."
    I would also like to thank my church families at Community Presbyterian
Church in Ventura, CA and North Coast Church in Carlsbad, CA and my
friends at Pierpont Bay Yacht Club. In particular, the Stephens Ministry
Program at CPC and North Coast Pastor Larry Osborne's book *Thriving in
Babylon*. Without the love and teaching of these dear friends who never left
my side during the darkest of times, I might never have endured. They were
often the only light in this very dark decade.

Special thanks to Lois Hart and Robyn E (My Cherokee Twin) and Travis Middleton who worked tirelessly without payment for now almost a decade trying to bring me justice in the legal system in Nevada detailing the crimes and obstruction of justice, which allowed the government to perpetrate these crimes against humanity.

I grieve the loss of colleagues and the many new friends who suffer from these devastating diseases and the deaths of their loved ones from ME/CFS, autism, and cancer. I know that no words can bring back the lost decades and loved ones but hope that this book can end the stigma and help heal their families from wounds too deep to imagine. Each and every one are my heroes and heroines. I cannot name them for fear they will face additional retribution.

From Kent:

I'd first like to thank my wonderful partner in life, Linda, and our two children Jacqueline and Ben, for their constant love and support. I'd like to thank my mother, Josephine, and my father, Jack, for teaching me to tell the truth regardless of the consequences and showing how to love through even difficult times. I'd like to thank the best brother in the world, Jay, and his wonderful wife, Andrea, and their three children, Anna, John, and Laura, for always being on my side.

I'd like to thank some of the wonderful teachers in my life, my seventh-grade science teacher, Paul Rago, my eighth-grade English teacher, Elizabeth White, my high school science teacher, Ed Balsdon, my religion teacher Brother Richard Orona, and in college, English professors Clinton Bond, Robert Haas, Carol Lashoff, and in the political science department, David Alvarez, who nominated me to be the school's Rhodes Scholar candidate. I'd also like to thank my college rowing coach, Giancarlo Trevisan, the mad Italian, who showed me what it means to have crazy passion for an often-overlooked sport. In law school, I'd like to thank Bernie Segal, the criminal defense attorney who taught me to always have hope that justice will eventually prevail. I'd like to thank my writing teachers, James Frey, who looked at me one time and said, "Yeah, I think you'll be a writer," as well as Donna Levin, and James Dalessandro, who always said to find the story first, then write the hell out of it.

My life wouldn't be complete without my great friends, John Wible, John Henry, Pete Klenow, Chris Sweeney, Suzanne Golibart, Gina Cioffi, Eric Holm, Susanne Brown, Rick Friedling, Max Swafford, Sherilyn Todd,

Rick and Robin Kreutzer, Christie and Joaquim Pereira, Tricia Mangiapane, and all of you who have made my passage through life such a party.

I work with the best group of science teachers at Gale Ranch Middle School, Danielle Pisa, Neelam Bhokani, Amelia Larson, Matt Lundberg, Katie Strube, Derek Augarten, and Arash Pakhdal. Thanks for always challenging my thinking and making me ask what is best for our students.

In the activist community, I'd like to thank J.B. Handley, Tim Bolen, Mary Holland, Lou Conte, Del Bigtree, Brian Hooker, Barry Segal, Elizabeth Horn, Brian Burrowes, Polly Tommey, Dr. Andrew Wakefield, and Robert F. Kennedy, Jr., for their friendship in the continuing fight against the Goliath of corrupted science.

Lastly, I'd like to thank my agent, Johanna Maaghul, my wonderful editors, Anna Wostenberg, and at Skyhorse the fabulous Caroline Russomanno, and for the faith shown in me over the years by publisher Tony Lyons.